LABORATORY MANUAL

ENGINEERING
OUR DIGITAL FUTURE

THE INFINITY PROJECTSM

Sally L. Wood • Marc P. Christensen
Scott C. Douglas • John R. Treichler

PEARSON

Prentice
Hall

Upper Saddle River, New Jersey 07458

Associate Editor: *Alice Dworkin*
Executive Managing Editor: *Vince O'Brien*
Managing Editor: *David A. George*
Production Editor: *Scott Disanno*
Supplement Cover Manager: *Daniel Sandin*
Manufacturing Buyer: *Ilene Kahn*

© 2004 by Pearson Education, Inc.
Pearson Prentice Hall
Pearson Education, Inc.
Upper Saddle River, NJ 07458

Printed in the United States of America
10 9 8 7 6 5 4 3 2 1

ISBN 0-13-035554-2

Pearson Education Ltd., *London*
Pearson Education Australia Pty. Ltd., *Sydney*
Pearson Education Singapore, Pte. Ltd.
Pearson Education North Asia Ltd., *Hong Kong*
Pearson Education Canada, Inc., *Toronto*
Pearson Educación de Mexico, S.A. de C.V.
Pearson Education—Japan, *Tokyo*
Pearson Education Malaysia, Pte. Ltd.
Pearson Education, Inc., *Upper Saddle River, New Jersey*

TABLE OF CONTENTS

Preface

The laboratory experiences and procedures described in this manual are designed to supplement the *Engineering Our Digital Future* textbook with hands-on interactive exploration of ideas and relationships from the text. Each laboratory from an Infinity Project Experiment box has a corresponding laboratory description with step-by step instructions. Although the individual laboratories on sounds or images or communications may focus on different topics, a common approach is used for all.

Students can quickly create a laboratory design, test it, and improve it using the Infinity Laboratory Kit. The kit includes a SPEEDY 33 circuit board which has a digital signal processor (DSP) chip and two small microphones on it. A camera is included for video input, and a set of speakers can be connected to the DSP board. The kit also includes the Visual Application Builder (VAB) software which is used to create designs and then run them.

The laboratories are based on a block diagram approach to problem solving that is used throughout the text. Big projects are broken up into several smaller functions or jobs, and each of these small jobs is individually more manageable than the one big job. Each small job is represented by a block, and blocks are connected to form the complete project or process. This approach works well on almost any large project whether it is an engineering design or planning a class picnic or producing a play. When all the small jobs are done well and the interactions between the small jobs are well planned, the larger project will be successful.

This block diagram approach to design is widely used in engineering practice. It is powerful and efficient because it allows us to concentrate on what we want the project to do at a functional level. The blocks take care of many of the details of implementation for us. At a later time we might want to look at the details of how an individual block does its job to make the project better, but we do not have to worry about all the details of all the blocks before we can start.

How to use this manual

A basic laboratory experience has been designed for each laboratory described in the *Engineering Our Digital Future* textbook. In most cases a laboratory uses an interactive Visual Application Builder (VAB) worksheet with the SPEEDY 33 digital signal processing circuit board and a PC. A few of the laboratories use other typical computer resources such as a media player application. The easy-to-use basic laboratories demonstrate ideas and relationships described in the text, and complete step-by step instructions are provided for each.

In some cases students and instructors may wish to use a basic laboratory as a starting platform for a more advanced project. Almost all of the basic laboratories have the potential to be extended in many directions depending on the amount of time available and the special interests of the class

At the end of some of the laboratories there are optional additional procedures called Quantitative Investigations or Further Explorations. These either use the worksheets in a more quantitative way or modify the worksheet to have additional capabilities. These optional sections may be used as written or they may be skipped without loss of continuity. And like the basic laboratories, they may be extended even further for special class projects.

The laboratories follow the development of the material in the textbook and also grow in design sophistication as progress is made through the chapters. The first chapter of the text contains a broad overview to the rest of the book, and there are no specific laboratory experiences directly related to it. Instead, one of the simple installation test laboratories is used to demonstrate the basic techniques needed to modify worksheets and then run them. Chapters 2 to 5 focus on designs that convert audio and image data to numbers and then analyze them or improve them or use them to create new audio and image data. In Chapters 6 to 8 more complex blocks are used for coding and communication.

An important benefit of these laboratory experiences is that they provide a way to show concrete demonstrations, in the form of images or sound, of mathematical concepts that are often considered only in abstraction. For example, the order of operations or the application of the distributive law shown on paper may involve simply moving parentheses and combining factors. When operations are done in a different order on sound or images, it is much easier to see the result of the change. This helps build an intuitive understanding. The concepts of a function, a function of a function, and an inverse function are explored visually in Laboratory 4.2 in the form of a mapping from input image values to output image values. Concepts of digital data representation using a limited set of values are a basic part of most of the audio and image laboratories.

When using these laboratories it is important to remember that they are engineering design laboratories and that engineering design includes science, mathematics, value judgments, and trade-offs between competing demands. For example, a project specification might call for high quality and low cost. But reducing cost at some level may cause an unacceptable reduction in quality, and different people may have different ideas about what is unacceptable. These laboratories include a combinations of analytical questions about observations which do have a well defined answer, and judgments about balancing competing requirements. While there may be no "right answer" to the judgment questions, students should always be able to justify their answers. It is often the challenge of balancing competing demands that leads to creative new solutions.

Additional information about laboratories can be found at www.infinity-project.org.

Acknowledgement

The authors gratefully acknowledge the tireless efforts of John D. Norris for his technical support and Rosemary G. Aguilar for composition and formatting. Without their support this laboratory manual would not have been brought to fruition.

Introduction to Block Diagram Design

What is a block diagram and how is it used?

A block diagram tells us what small jobs or processes we need in order to get a larger project done. Each block has a specific function to perform. Each block also has specific set of inputs and outputs. By connecting the output of one block to the input of another, we can have the combined function of both blocks. This allows us to take simple elements that are individually easy to understand and combine them to achieve a more complex goal. The selection of the specific blocks and the way they are connected determines how the larger project will operate and how well it will work.

A block diagram describes all the small jobs that are needed, but it does not tell us exactly how to do each small job. For example, an engineer might decide that a particular block in a big project could be done best with a special custom designed circuit or with a general purpose computer program or with a mechanical device. The choice would depend on things like how well each of the methods works, how hard each is to do, and how much each would cost.

In these laboratories our blocks will create instructions for the digital signal processing (DSP) circuit on the SPEEDY 33 board or for the PC. Both the PC and the external DSP board have very fast processors for computation, but they are designed to work best in different circumstances. The PC is the better choice for interaction with the user through the keyboard and mouse and display screen. It also can hold large amounts of data in the memory on its disk. In contrast, the DSP board is better suited for handling audio signals and producing outputs immediately without delays that are normally associated with PCs and PC software. In some cases, such as laboratories using images, the blocks in our block diagram will use the PC. In other cases, for example when audio signals are used, the blocks will use the SPEEDY 33 DSP board. Although we normally keep the DSP board connected to the PC, it is possible to load instructions into the DSP board and then disconnect it so that it can run independently from the computer. This is important when the project must be small and mobile like a cellular telephone or most robotics applications.

How does a block diagram make instructions for the PC or the DSP board?

The Visual Application Builder (VAB) software allows us to quickly make block diagrams on a computer screen worksheet with a simple graphical editing method. We can select blocks from a library that do the jobs we need and then place these blocks on the worksheet. After we have connected the inputs and outputs of the blocks on the worksheet, we have a visual block diagram of our complete process. The VAB automatically converts the blocks from our block diagram into instructions for the PC or the DSP board. When we start our worksheet, the PC and DSP board will execute those instructions and perform the functions our block diagram describes.

Suppose, for example, we want to simply connect a sound input from a microphone to a set of speakers so we could hear the sound. First we would select a microphone block and a speaker block from the library and add them to our worksheet. Then we would draw a line from the microphone output to the speaker input to make the connection. That would complete our design. After starting the worksheet, whatever sound was picked up by the microphone would be heard coming from the speakers. If we wanted to add sounds together or change the sound from the microphone, we would put more blocks between the microphone and speaker.

Blocks used in these laboratories

Every block has a well defined function or job. It also has a well defined set of inputs and outputs. For example, a block might have two inputs, a number and an image. Sometimes blocks also have parameters or control inputs which can change the function of a block. A block could be as simple as an adder block which adds its two inputs together to create an output sum. We can have an adder block for sounds or an adder block for images. A block can also perform a complex function such as analyzing an image sequence to find the locations where there is motion.

Most projects that are interesting respond to the world we experience and cause something to happen in the world we experience. So in addition to simple and complex processing blocks, we also need input and output blocks. These blocks are very important because they transfer audio and image information between the DSP board and the outside world.

Input Blocks: Blocks which have no inputs on the worksheet represent input devices. That means that the block input comes from outside VAB as described below. The VAB has a variety of input blocks.

- A microphone block has an input from the sound in the air around it. The block includes both the connection to the physical microphone, which converts the pressure variations in the air around it to electrical signals, and the electronic circuit on the DSP board which converts the electrical signal into a sequence of numbers. The microphone block creates an output signal, which is a sequence of numbers, for use as an input by other blocks in a worksheet.
- A camera is also an input device. Its input comes from the light that falls on the camera's sensors, and its output is a set of numbers defining the content of an image.
- A block that reads from a stored data file can be used as an input block for sounds and images.
- A keyboard gets its input from the keys you type. Its output is a string of characters that can be used by other blocks
- A block can also get its input from a mathematical operation. For example, an input block can use a cosine function to create a cosine signal for other blocks to use.

Output Blocks: Blocks which have no outputs on the worksheet represent output devices. That means that whatever output the block produces goes outside VAB as described below. The VAB also has many different kinds of output blocks.

- A speaker is an example of an output device. Its input is a signal or sequence of numbers from the output of some block on the worksheet. Its output is the sound we hear from the speakers. The block includes both the circuit which converts a sequence of numbers into an electrical signal and the connection to the physical speakers that produce the sound from the electrical signal.
- Visual output blocks include image displays, plots of graphs, and numeric or text displays.

Simple rules for connecting blocks: Blocks can be connected to build a worksheet using a few simple rules.

- Wires drawn between blocks represent information that is produced by one block and used by the other. A wire can carry an audio signal or an image or a simple number. This means that the input of a block should be connected to the output of another block that is of the "same type."

- Only one wire can be connected to each input of a block.
- The output of a block can be connected to inputs of many blocks.

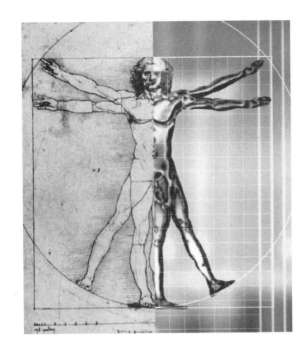

The World Of Modern Engineering

Chapter 1 introduces students to the engineering design process and the basics of modern technology including integrated circuits, computer chips, mathematical concepts such as Moore's Law, binary numbers, and simple exponential functions that describe constant growth rates.

Infinity Labs

1.1   Testing your system

1.2   High Tech Demos

Introduction

The purpose of the laboratories in Chapter 1 is to introduce the VAB and Speedy33 DSP board both at the high level of interesting projects and also at the basic operational level. Two future laboratories are previewed in the High Tech demos section. The Echo Generator is fully described as Lab 2.9 and the Object Tracker can be found in the Motion Detector Lab 4.9. These laboratories are relatively intuitive and can be used to demonstrate the capabilities of the block diagram design technique used with the DSP board and the PC.

One of the Testing Your System laboratories is also used here to demonstrate how to use VAB. The laboratory procedure provides detailed instructions for basic VAB operations that will be used in future laboratories.

1.1 Testing Your System

Lab Objectives

This laboratory will help you become familiar with the use of VAB and the Speedy 33. You will listen to audio from a microphone that has been combined with audio created by the Speedy 33 processor. You will be able to view a plot of the sound on the worksheet display screen. You will learn much more about audio signals and cosines in the laboratories in Chapter 2. In this laboratory you will learn how to connect blocks and how to change what the blocks do. More details about using worksheets can be found in the VAB tutorial. To view the tutorial, click the **Help** button on the top menu bar.

Textbook Reading

*   Prerequisite textbook reading before completing this lab: pp. **7-11** and **23-27**.

Engineering Designs and Resources

Worksheets used in this lab:

*   **L01-00-01 Testing Your System Mic Cosine.Lst**: Assures that the microphone in the Infinity kit is working as expected
*   **L01-00-02 Testing Your System Camera.Lst:** Assures that the camera connected to the PC is working as expected.

1.1.1 Testing Your System with Microphone and Cosine

Worksheet Description

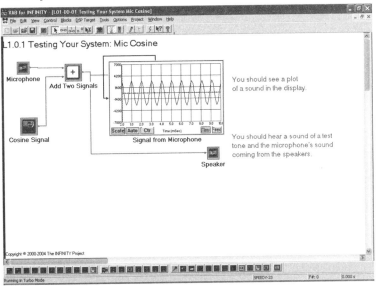

Figure 1.1 Testing Your System - Microphone and Cosine

When you open the worksheet for this laboratory, you should see a display that looks like the Figure 1.1 above except that no plot will be shown on the Signal from the Microphone display. On the left side of the worksheet there are two blocks that make audio signals. The Microphone block brings the sound from the microphone into the SPEEDY 33 circuit board. The Cosine Signal block creates a single tone signal on the circuit card

An adder block adds the two audio signals. The two inputs to the adder block are connected to the output of the Microphone block and the Cosine Signal block. These connections are shown with red wires to indicate that they are in the external SPEEDY 33 circuit. The output of the adder is connected by a red wire to a Speaker block. This lets us hear the sound from the microphone and the single tone on the speakers which are directly connected to the circuit board.

In order to see the plot of the sound on the worksheet screen, we use a display block labeled Signal from Microphone. Note that the input to this block is a white wire indicating that it is inside the computer which is running VAB. Underneath the display block there is a block which takes signals from the Speedy 33 circuit and sends them to the PC. This block's input is a red wire connected to the adder block output.

In all of the VAB laboratories we will use a graphical editor to create, modify, and operate worksheets. A summary of the most often used techniques and their associated icons is shown here. These will be described in more detail in the steps of the following laboratory procedure.

Summary of Often Used VAB icons

1. Setup tool

2. Connection tool

3. Conditional Connection Tool

4. Parameter Connection Tool

5. Delete Tool

6. Start the worksheet (green light)

7. Stop the worksheet (red light)

Laboratory Procedures

Steps	Instructions
Step 1:	Open the worksheet *L01-00-01 Testing Your System Mic Cosine.Lst*. Do this by using the left mouse button to click **File** on the left side of the top menu bar. Then select **Open** from the drop down menu. • Use the Browser to navigate to Program Files/Hyperception/VABINF/Labs/ CH01/L01-00 Testing Your System/. • Select the file *L01-00-01 Testing Your System Mic Cosine.Lst*. • Click **Open** in the lower right hand corner of the browser window. You should see a worksheet similar to the one shown in the figure above, but with no plot shown on the Signal from Microphone display. Before you start the worksheet, make sure that the speakers are some distance away from the microphones on the SPEEDY 33 circuit.
Step 2:	Start the worksheet by using the left mouse button to click the **green traffic light** icon on the horizontal icon tool bar just below the top menu bar. • You should see a plot similar to the plot shown in the figure above. It might be shifted to the left or the right. • You should hear a tone coming from the speakers connected to the circuit board. Adjust the speaker volume to a low level that is comfortable. If you do not hear a tone, make sure that speakers are connected the circuit card at the OUT/Speakers connector. Make sure that the speakers are turned on.
Step 3:	Observe the plot of the sound.
	Q1: What is the time duration shown on the horizontal axis?
	Q2: How many positive maximum points do you see on the plot? How many negative minimum points do you see on the plot? How many times does the plot change from positive to negative by crossing the horizontal axis?

Steps	Instructions (Continued)
Step 4:	We can change the sound of the tone by changing one of the parameters of the Cosine generator. The menu shown below will appear after either right clicking the **Cosine Signal** block or double left clicking it. We see that the amplitude is 3,000 and the frequency is 1,000. Change the frequency to 500 by clicking the 1000.0 entry and typing 500 in its place. Then click **OK**. **Figure 1.2** DSP Cosine Generator Parameters window
Step 5:	The display of the sound signal should now look like the figure shown below. Observe how this display is different from the previous display. **Figure 1.3** Worksheet after cosine frequency change.

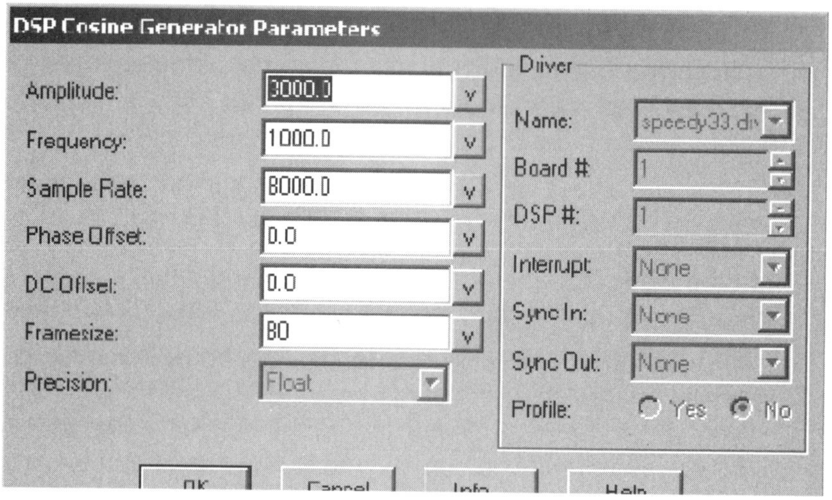

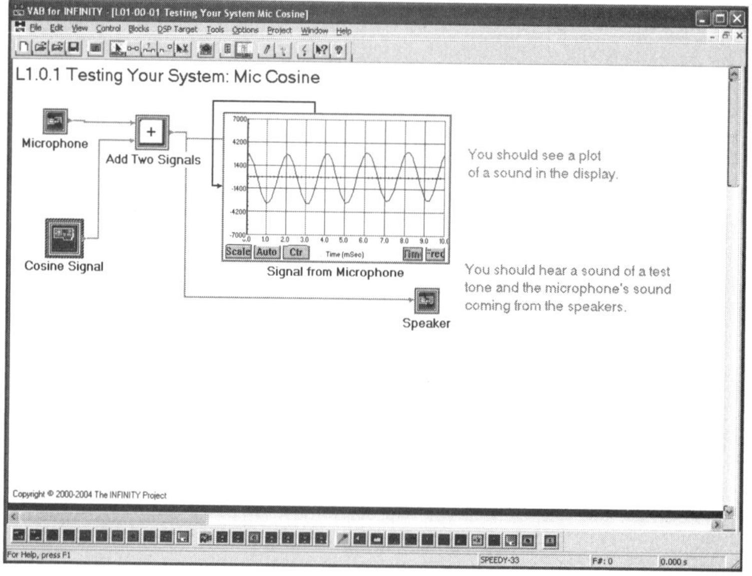

Steps	Instructions (Continued)
	Q3: Now how many positive maximum points do you see on the plot? How many negative minimum points do you see on the plot? How many times does the plot change from positive to negative by crossing the horizontal axis?
Step 6:	Talk or whistle into the microphones on the circuit board. The display should change to show the sound of your voice added to the tone. Both the speech sound and the tone should be heard from the speakers. 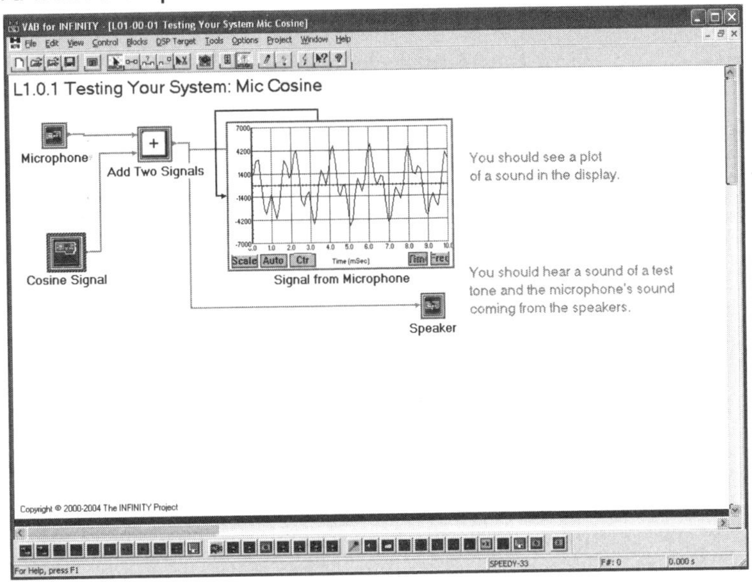 **Figure 1.4** Worksheet with display of whistle and cosine.
Step 7:	We can make even more interesting sounds if we multiply the two audio signals instead of adding them. To change the worksheet to do multiplication instead of adding, we will have to add a new block and make some new wire connections. These can all be done easily using the tools on the menu bar.
Step 8:	Stop the worksheet. This can be done in two different ways: • Use the left mouse button to click the **red traffic light** icon on the horizontal icon tool bar just below the top menu bar. Or: • Hit the **Esc** key on the keyboard.

Steps	Instructions (Continued)
Step 9:	Add a block to multiply two signals: Click **Blocks** on the top menu bar. From the drop-down menu click **Select Blocks**. You should see the pop-up menu shown below. • Select the **Infinity Hot List** for the library. • For the Group List on the left select **DSP Blocks**. • In the Function List on the right click **DSP Multiply**. • Then click the **Add to Worksheet** button. • Click the **Close** button to close the pop-up window. You should now see the new block on your worksheet. Left click it to select it, and drag it to a position just below the adder block. 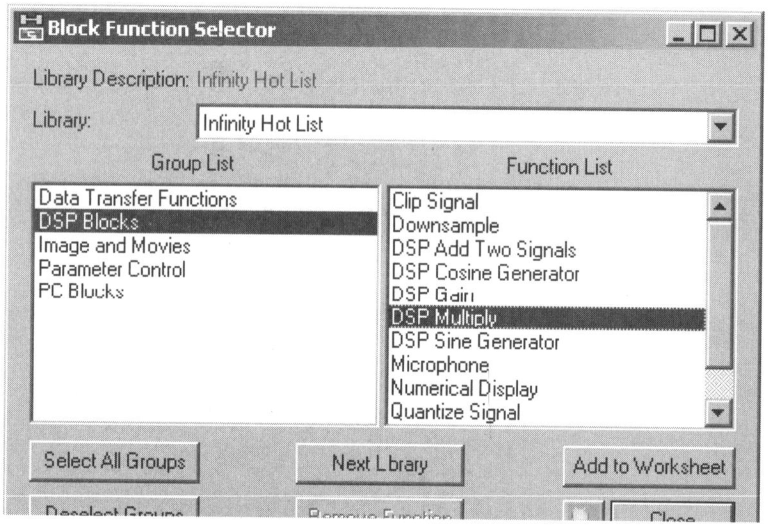 **Block Function Selector** Library Description: Infinity Hot List Library: Infinity Hot List Group List Data Transfer Functions DSP Blocks Image and Movies Parameter Control PC Blocks Function List Clip Signal Downsample DSP Add Two Signals DSP Cosine Generator DSP Gain DSP Multiply DSP Sine Generator Microphone Numerical Display Quantize Signal Select All Groups Next Library Add to Worksheet Deselect Groups Remove Function Close **Figure 1.5** Block Function Selector window
Step 10:	Now we have to disconnect the adder from the speaker and display. But first we have to move the Signal from Microphone display. • Left click the display to select it. • Then drag it to the right to reveal the hidden **Upload to Computer** block.
Step 11:	Click the deletion tool on the horizontal icon bar. (It looks like a pair of scissors.) • Click the red wire from the adder to the speaker block to delete the wire. • Click the red wire from the adder to the Upload to Computer block to delete the wire.

Steps	Instructions (Continued)

Step 12:	Connect the new multiplier block:
	
	Click the connection tool icon at the top of the worksheet. (It looks like a horizontal line connecting two points.)
	
	• Connect the microphone output to the upper input of the DSP Multiply block.
	    Left click inside the microphone block near the output connector on the right side of the block.
	    Then left click inside multiply block near the top input connector.
	    You should see a new wire connecting the two blocks.
	• Connect the Cosine output to the lower input of the DSP Multiply block.
	• Connect the output of the DSP multiply block to the input of the speaker block.
	• Connect the output of the DSP multiply block to the input of the Upload to Computer block.
	
	Now your worksheet should look like Figure 1.6 below.
	
	Figure 1.6 Modified worksheet

Step 13:	After making all the connections, click on the **setup tool** at the top of the worksheet. (It is the diagonal arrow next to the connection tool.)
Step 14:	Change the parameters of the cosine block because we are multiplying now instead of adding. Right click the Cosine Generator block as we did in Step 4. • Change the frequency to 1000.0 • Change the amplitude to 5. • Click **OK**.
Step 15:	Start your worksheet. Speak or whistle into the microphones on the circuit board. You should hear a sound from the speakers that makes your voice sound like a robot or a space alien in movies.

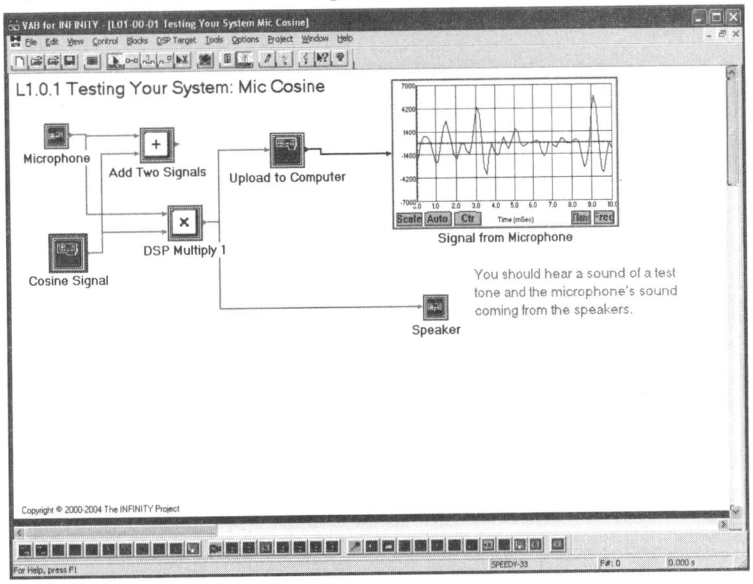

Steps	Instructions (Continued)
Step 16:	Stop the worksheet.
	Do this by using the left mouse button to click **File** on the left side of the top menu bar. Then select **Close** from the drop down menu.
Step 17:	Add a slider control so that we can change the frequency without going through the procedure of Step 4 every time we want to hear a new frequency.
	As we did in Step 9, click **Blocks** on the top menu bar. From the dropdown menu click on **Select Blocks**. You should see the pop-up menu shown below.
	• Select the **Infinity Hot List** for the library. • For the Group List on the left select **Parameter Control**. • In the Function List on the right click **Slider Bar Control**. • Then click the **Add to Worksheet** button. • Click the **Close** button to close the pop-up window. Now you should have a slider block on your worksheet.
	Right click the upper left corner and drag the icon to a position below the cosine generator. (You may also want to make it taller and thinner using the arrow adjustments on the sides of the block.)
Figure 1.7 Block Function Selector window |

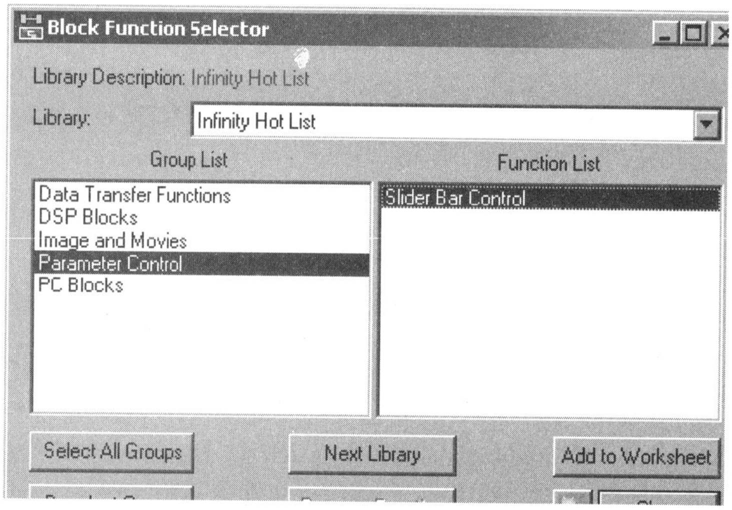

Steps	Instructions (Continued)
Step 18:	Set the parameters of the slider. Right click the slider. You should see a popup menu like the one shown below. Change the following parameters. • Top value: 4000 • Bottom value: 0 • Number of steps 401 Then click **OK**. The slider will allow you to set frequency values that are multiples of 10. 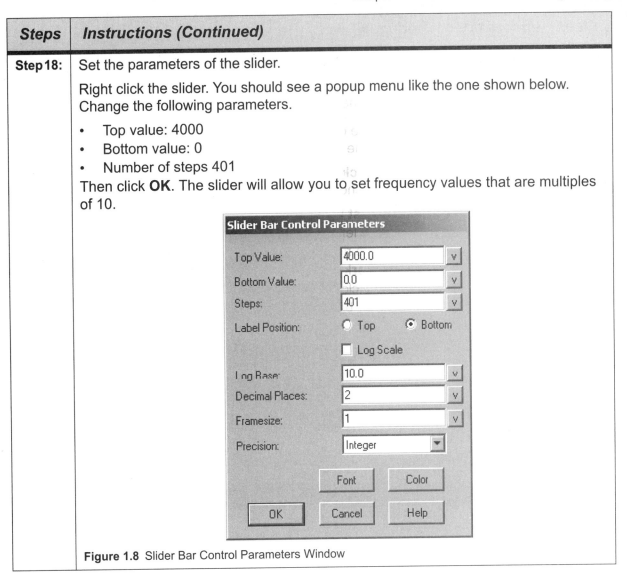 **Figure 1.8** Slider Bar Control Parameters Window

Steps	Instructions (Continued)
Step 19:	Select the **parameter connect** icon from the horizontal menu of icons. (It looks similar to the connection icon except that it has a line with a reflected "L" shape with two "points" at each end. It is next to the deletion icon with the scissors image.) Left click the slider block. Then click inside the Cosine generator block. You should see a pop up menu like the one shown below, which lets you choose which parameter you want to control. Select **Frequency** since that is the parameter we want to control with the slider. Then click **OK**. 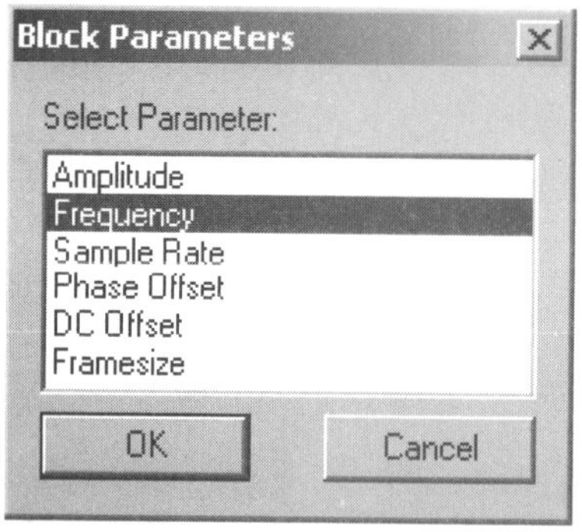 **Block Parameters** ☒ Select Parameter: Amplitude Frequency Sample Rate Phase Offset DC Offset Framesize [OK] [Cancel] **Figure 1.9** Making a parameter connection
Step 20:	Click the setup tool at the top of the worksheet. (It is the diagonal arrow next to the connection tool.)
Step 21:	Start your worksheet. Speak into the microphone as you adjust the frequency of the cosine and listen to the results.
	Q4: What happens to your voice with the slider set to create low frequencies?
	Q5: What happens to your voice with the slider set to create high frequencies?
	Q6: What frequency gives the most interesting effect?

Steps	Instructions (Continued)
Step 22:	Stop the worksheet.
	Close the worksheet.
	Do this by using the left mouse button to click **File** on the left side of the top menu bar. Then select **Close** from the drop down menu.

1.1.2   Testing Your System with a Camera

Worksheet Description

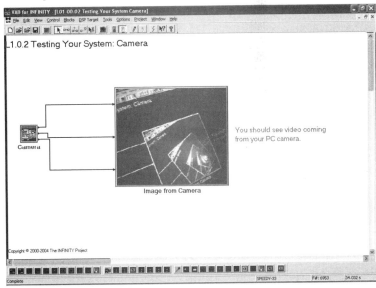

Figure 1.10 Testing Your System Camera

When you open the worksheet, you should see a display that looks like the Figure 1.10 above except that the image display will be black. This simple worksheet has only two blocks. The input block is a camera and the output block is a color image display. The three wires that connect the camera to the display have the red, green, and blue components of the color image. Note that the wires are white. This indicates that the connections occur inside the PC, not in the SPEEDY 33 DSP board.

Using the procedures described in the previous laboratory, open the worksheet *L01-00-02 Testing Your System Camera* and start the worksheet. You can point the camera at objects in the room and see them on this display.

Stop the worksheet and close it as we did in the previous lab.

This laboratory and the previous laboratory can be used as basic test laboratories for the microphone, the speakers, and the camera because the worksheets have so few blocks.

1.2 High-tech Demos

Lab Objectives

The previous two laboratories demonstrated very simple worksheets. These two laboratories provide a preview of the capabilities of more complex worksheets with audio and video inputs. You can try them out now and explore what they can do. Later in Chapters 2 and 4 you will see these laboratories again and do more with them.

Textbook Reading

- This lab appears on page **27** of the Engineering Our Digital Future textbook.
- Pre-requisite textbook reading before completing this lab: pp. **4-26**.

Engineering Designs and Resources

Worksheets used in this lab:

- **L01-01-01 High Tech Demos Echo.Lst**: Allows students to control the strength and delay of an echo.
- **L01-01-02 High Tech Demos Object Tracker.Lst**: Allows students to detect motion of objects in a video sequence.

1.2.1 High Tech Demos - Echo

Worksheet Description

Echoes are important for creating sound effects. They are also important for medical tests and exploring for oil. This laboratory lets you create an echo of whatever sound is picked up by the microphone. Normally an echo is created when sound bounces off a distant wall and comes back so you hear the same sound again after some delay. We can create the same effect with a digital delay.

When you open the worksheet you should see a screen that looks like the figure below. The microphone on the left side responds sounds and creates sound signals on the DSP board. The echo of the sound signal is created by the Delay block. Then the sound signal from the microphone and the delayed sound signal are added together with an Add block. The sum of the sound signals is connected to the speaker block, which converts the signal on the DSP board into sound that we hear.

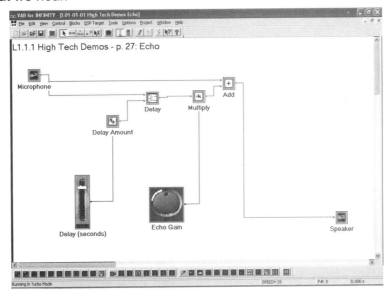

Figure 1.11 High Tech Demos - Echo

You can control the echo in two ways. You can adjust the Delay slider to make a delay time of up to half a second. You can also control the strength of the echo with the Echo Gain knob.

Open the worksheet *L01-01-01 High Tech Demos Echo.Lst* and start the worksheet. If the speakers immediately start to make a loud tone, move them away from the microphones on the board so there is no feedback.

Make some sharp sounds near the microphone by clapping your hands or snapping your fingers and listen to the echo. Adjust the delay and echo gain while you are doing this. Try speaking into the microphone and adjusting the delay and echo gain while you are talking.

Stop the worksheet and close it as we did in the previous lab.

1.2.2   High Tech Demos - Object Tracker

Worksheet Description

An object tracker could be useful in a wide variety of applications from detecting intruders to counting animals in the wild. In the previous laboratory we used a sound delay to create an echo. In this laboratory we will use an image delay to detect moving objects in the camera's field of view.

When you open the worksheet you should see a screen that looks like the Figure 1.10 below except that the image displays will be blank. The camera and the Original Image display look exactly like Lab 1.0.2. Here the image from the camera also goes to an Image Delay block, and then a subtract block takes the difference between the original image and the delayed image. This is shown in the Difference Image display. The difference image is also processed by an

Object Tracker block and recombined with the original image to highlight moving parts. Green lines show the horizontal and vertical positions of moving objects.

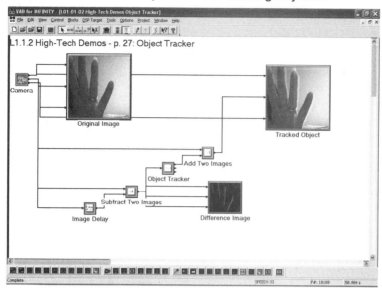

Figure 1.12 High Tech Demos - Object Tracker

Open the worksheet *L01-01-02 High Tech Demos Object Tracker.Lst* and start the worksheet. Aim the camera at a scene where nothing is moving and see what happens. Then move something in front of the camera and watch the motion being detected. Try aiming the camera at a hallway where people are walking by and watch the object tracker follow them.

Stop the worksheet and close it as we did in the previous lab.

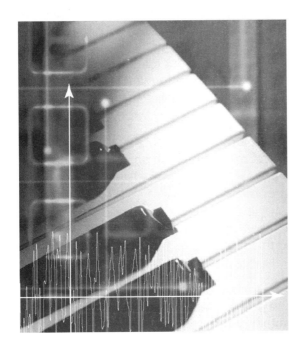

Creating Digital Music

Chapter 2 exposes students to some of the most important engineering ideas associated with the creation of digital music. Students learn how basic ideas drawn from the right triangle such as sines and cosines are fundamental to making computer music.

Infinity Labs

Introduction

The laboratories in Chapter 2 experiment with sounds, which are audio signals. Audio signals are obviously important for speech and music, but they are also important for communications systems and many other applications. Signals from speech or music or mathematical computations can be changed or combined, and the results can be viewed as mathematical plots or listened to as sound from speakers.

The VAB blocks that work with sounds use both the DSP board and the computer. The most basic input block is a microphone block, which converts the pressure variation in the air that we hear as sound into a sequence of numbers that we can use for computations. The corresponding basic output block is the speaker block, which does the opposite conversion from a sequence of numbers into sound. Some laboratories use the microphone on the DSP board and others may use the microphone on the computer, so VAB has a block for each. There are also separate VAB blocks for the speakers connected to the computer and the speakers connected directly to the DSP board.

Sound input may also come from files stored on the computer as *.wav* files or from mathematical computation, which often uses sines and cosines. These sounds can be combined with sounds from the microphone using a variety of VAB blocks. Another commonly used output block is a plot of the sound signal as a function of time. The VAB display blocks that plot sound use the computer, so signals from the DSP board must be sent to the computer to be plotted on the worksheet. Note that in the worksheets red lines connect blocks using the DSP board and white lines connect blocks using the computer. Often in music or communications the frequency or the pitch of a sound is important, so there are VAB blocks which also display this information.

2.1 Plots of Speech

Lab Objectives

The objective of this lab is to see and explore the shape and form of your own voice, the sound of music, and other common sounds.

Textbook Reading

*   This lab appears on page **44** of the *Engineering Our Digital Future* textbook.
*   Prerequisite textbook reading before completing this lab: pp. **33-44.**

Engineering Designs and Resources

Worksheet used in this lab:

*   **L02-01-01 Plots of Speech.Lst:** This worksheet allows you to see sound signals on your computer screen.

Worksheet Description

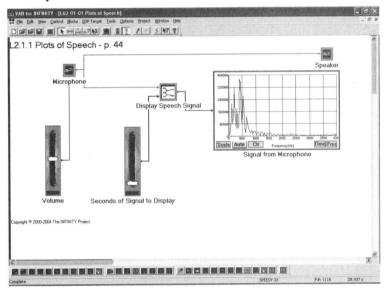

Figure 2.1 Plots of Speech

Open the worksheet *L02-01-01 Plots of Speech.Lst*. You should see something similar to Figure 2.1 above.

*   **Microphone** – This block takes in a signal from the microphones on your DSP board.
*   **Volume** – The Volume Slider block adjusts the gain of the microphone signal by multiplying it by a constant.
*   **1 Channel Display with Buffer** – This block collects consecutive values from the signal it is fed so that they can be displayed on the PC in front of you.
*   **Seconds of Signal to Display** – This block is a slider that controls how much of the signal you will see in the Display on your computer
*   **Signal from Microphone** – This block displays the signal from the microphone.
*   The blocks are connected together such that all signals flow from the left to the right.

Laboratory Procedures

Steps	Instructions
Step 1:	Start the worksheet.
Step 2:	You should start seeing a signal being displayed on the Signal from Microphone display.
	The display has two axes; the x-axis is time, and the y-axis is the value of the signal.
	Q1: What do you notice? Is the signal moving a little or a lot?
Step 3:	Try speaking into the microphone.
	Q2: What happens?
Step 4:	The display doesn't show every part of the signal all the time; only the parts of the signal that fall within the y-axis range can be shown. Sometimes the signal is so large that it falls outside of this range.
	So, to get the y-axis to scale with the size of your microphone signal, click on **Auto** until it is greyed-out. Now, the y-axis should change depending on how loud the signal is.
	Talk loudly into the microphone.
	Q3: What happens to the scale?
Step 5:	Usually, we want a fixed y-scale, so to get a good scale for your voice, say "Ahhhhh" for a long time into the microphone while simultaneously clicking on **Scale**.
	Now, the y-axis limits should be fixed at a large value. When you talk into the microphone again, it should show both the loud and soft parts of your voice.

Steps	Instructions (Continued)
Step 6:	Another thing you can do with this worksheet is capture your voice as if it is frozen in time. To do this, halt the worksheet while you are talking into the microphone. You should see the part of your voice that the system recorded just before you stopped the worksheet. Move your mouse over the display near any part of the signal – it should give you the value of the signal and the time that it occurred. (Alternatively, click on the worksheet near the portion of the signal you want to measure.)
Step 7:	Now we are ready to do some experiments with your *Plots of Speech* worksheet. The first thing we will do is look at your voice over different time scales. The second slider on your worksheet allows you to control how much of your microphone signal is displayed. It should be set at its minimum value of 0.005 seconds. If not, move the slider to this value. Now, slowly move the slider up.
	Q4: What is happening on the display?
	Q5: What do you notice about the signal?
Step 8:	Move the slider to its top-most value. Now, talk into the microphone.
	Q6: Can you see individual words in the display?
Step 9:	Count "1, 2, 3..." in to the microphone to see the distinct words to be sure. You have total control as to how much of the microphone signal you can see. Adjust the slider until it shows a value of 1 second, and count using "one one-thousand, two one-thousand,.." and so on. Each count should roughly fit into a single display time.

Steps	Instructions (Continued)
Step 10:	Now, we are going to "zoom in" to your voice signal to see its fine structure. Adjust the Seconds slider to its minimum value of 0.005, say "ahhhh" as if you are having your throat examined.
	Q7: What do you see on the display?
Step 11:	Try saying "Ooooh" in a high-pitched voice.
	Q8: What is happening now?
Step 12:	You should see the signal repeat over and over. Signals that repeat over and over are called periodic signals. Musical instrument sounds are periodic signals, too.
Step 13:	To make a really clean periodic signal, whistle into the microphone.
	Q9: What do you see?

Steps	Instructions (Continued)
Step 14:	While whistling into the microphone, halt the worksheet to freeze the signal in the display.

The signal should look something like the display below. Notice how smooth the signal is and how it repeats over and over. The time it takes for the signal to repeat is called its period.

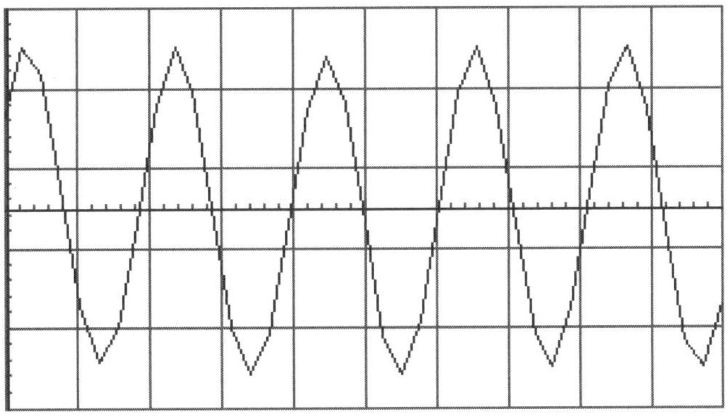

Figure 2.2 Signal for a whistle sound

Using the mouse, calculate the period of the signal in the following way:

1. Move the mouse near a peak of the signal.
2. Write down the x-axis (time) and y-axis (amplitude) values near this point.
3. Move the mouse near the next peak of the signal in the display to the left of the value you just recorded.
4. Write down the x-axis (time) and y-axis (amplitude) values near this point. The y-axis value should be close to that you measured above.
5. Subtract the second value from the first.

First Value	
Second Value	
Difference	

You should get a number in milliseconds.

Q10: What is the period of your whistle?

Another parameter of a periodic signal is its amplitude. The amplitude of the signal is the largest value away from zero that the signal has.

Steps	Instructions (Continued)
	Q11: What is the amplitude of your whistle?
	Yet another parameter of a periodic signal is its fundamental frequency. The fundamental frequency of a periodic signal is related to its period T by $$f = 1/T$$
	Q12: What is the fundamental frequency of your whistle?

Steps	Instructions (Continued)
Step 15:	The fundamental frequency is what we recognize in music as the pitch of an instrument. By making signals that are periodic, we make music.

While it is somewhat of a burden to calculate the fundamental frequency of a signal using the period, it is the most accurate way to find its value. But there's another way: The Display has a button, called **Freq**, that calculates the spectrum of the signal that is fed to it. The spectrum displays all of the frequency components of one signal at one time. The x-axis of this display is in Hertz [Hz], and the y-axis is amplitude.

Click **Freq** with your frozen whistle signal. You should see something similar to Figure 2.3 below.

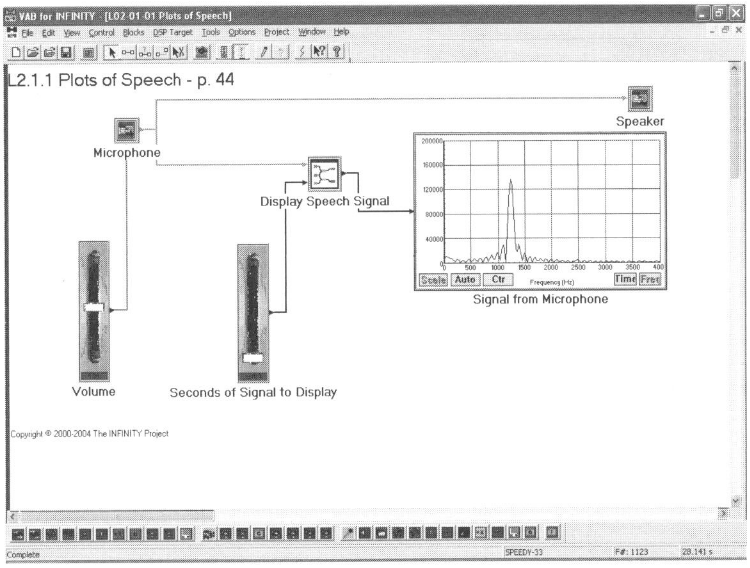

Figure 2.3 Plots of speech - Frequency

Notice how there's one large peak in the spectrum. For signals that are particularly simple such as your whistle, the peak of the signal should correspond to the fundamental frequency.

Use your mouse to find out the frequency of this peak in your display. |
| | **Q13:** Is it near the value that you calculated above for the fundamental frequency? |
| | **Q14:** What do you see in your plot? Does it look like Figure 2.3? |

Steps	Instructions (Continued)
Step 16:	Now, run the worksheet again, and whistle into the microphone. Leave the display showing Frequency. See what happens as you whistle lower and then higher. If you can't whistle, say "oooh" in as steady a voice as you can manage.
	Q15: What is the highest fundamental frequency that you can whistle?
	Q16: What about the lowest?

Overview Questions

A: What is the period of the signal? How is it related to fundamental frequency?

B: What are the units of fundamental frequency?

C: What is a typical period of your voice? What about a whistle? How are these related to how high or low the fundamental frequencies are for these signals?

Summary

This lab has helped you take apart your voice and analyze its structure using the concepts of period and fundamental frequency. How can we use these ideas to make music? The labs later in this chapter help you do that.

2.2 Generating Sine and Cosine Signals

Lab Objectives

The objective of this lab is to show how sines and cosines can be used to make interesting and useful sounds. You are probably familiar with the sin(A) and cos(A) keys on a calculator. You probably also know that sines and cosines are connected to right triangles--but did you know that sin(A) and cos(A) can also be used to make the pure tones of simple music?

Textbook Reading

*   This lab appears on page **65** of the *Engineering Our Digital Future* textbook.
*   Prerequisite textbook reading before completing this lab: pp. **44-65.**

Additional Materials

*   Scientific Calculator

Engineering Designs and Resources

Worksheet used in this lab:

*   **L02-02-01 Generating Sine and Cosine Signals.Lst**: A virtual experiment in making and listening to sines and cosines.

Worksheet Description

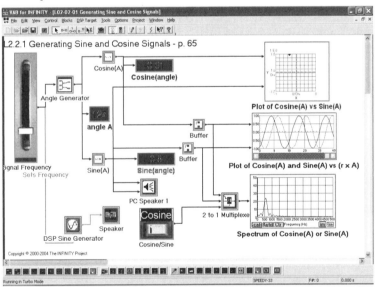

Figure 2.4 Generating Sines and Cosines

Open the worksheet *L02-02-01 Generating Sine and Cosine Signals.Lst.*

On the left side of this worksheet is a single slider that controls the rotational speed for the generation of the signals and the displays on the right side of the worksheet.

The angle generator block outputs an angle (between 0 and 359.9 degrees) as a function of time. Its input is the rotational speed which comes from the slider. So, we get numbers shown in the "angle A" display that go from 0 slowly up to 359.9 and then instantaneously go back to zero and start over again.

The fundamental frequency of a sound signal is the mathematical quantity that we associate with how "high" or "low" the pitch of a signal is. It is directly related to the period or repeating interval of the signal by the relation $f = 1/T$, where the period T is measured in seconds. The units of frequency are Hertz, or Hz for short.

The sin(A) and cos(A) blocks just evaluate the sine and cosine functions for the current value of x. We then do three things with these sin(A) and cos(A) values:

We display them on an x-y plot, with sin(A) controlling the x-axis and cos(A) controlling the y-axis.

We store the sin(A) and cos(A) values in a buffer that is like a digital recorder, keeping track of the most-recent values of these functions. The recorded signal is then sent to a display that shows the shapes of these functions over time.

The bottom plot shows the frequency content of the signals. Which function it is looking at depends on the position of the pushbutton switch next to it. This will be discussed further later.

The third display gives you a way to measure the fundamental frequency of the sinusoid.

Finally, there are two blocks near the bottom of the worksheet that run on the DSP processor board. Their job is to create and play an audio signal like the one that the sin(A) function is making on the screen.

Laboratory Procedures

Steps	Instructions
Step 1:	Start the worksheet.
Step 2:	The displays on the right side of the worksheet should begin to change.
	Q1: What do you see in the first display?
	Q2: What is happening to the white dot?

Steps	Instructions (Continued)
Step 3:	As we explained in the Instructions section, the x-axis of this first display is being controlled by the sin(A) signal path, whereas the y-axis is being controlled by the cos(A) signal path. Stop the worksheet and write down the value in the "Angle" display. Then, using a calculator, calculate sin(A) and cos(A) from this number <table><tr><td>**Angle**</td><td>**Sin(A)**</td><td>**cos(A)**</td></tr><tr><td></td><td></td><td></td></tr></table>
	Q3: Do the sin(A) and cos(A) values match the x-axis and y-axis values of the white dot from this first plot?
Step 4:	Start the worksheet again and halt it after a short time. The "Angle" value should be different now. Repeat the above calculations for this new angle and record the information on the chart above in Step 3.
	Q4: Do the values match up in this case?
Step 5:	Let's look at the second plot. It shows the sin(A) and cos(A) functions over time.
	Q5: What is happening to the functions on this plot?

Steps	Instructions (Continued)
	Q6: How are they related to the motion of the white dot in the display above? Slow down the dot by moving the Signal Frequency slider to a value of 30 to make it easier to see.
	Q7: Which color is the cos(A) plot--the yellow line or the red one? How can you tell? Start and stop the worksheet several times to check.
Step 6:	Now we are ready to do some experiments with this worksheet. The slider gives you a special way to control the worksheet. Change the value of the slider so that it reads 750 and run the worksheet.
	Q8: What is happening to the white dot in the first display?
	Q9: How about the functions in the second display?
	Q10: How are these behaviors different from what you first saw when the slider gave a value of 500? How is the sound different?
Step 7:	Set the value of the rotating slider to 750, and listen to the resulting sound signal that is produced. Then, change the slider value to 1250.

Steps	Instructions (Continued)
	Q11: What happened to the sound signal?
Step 8:	Finally, look at the bottom display on the right as you are running the worksheet. It should have the shape of a single peak, because sine and cosine signals are functions of a single frequency. If it doesn't, click on the **Freq** button on the bottom of the display.
	Q12: How is the position of this peak related to the slider setting?
Step 9:	Starting from the "Cos" toggle button position, switch the toggle button position from **Cos** to **Sin** and back again.
	Q13: What happens to the bottom plot?
	Q14: Does this make sense to you?
Step 10:	Stop the worksheet. Close the worksheet.

Overview Questions

A: How are the functions cos(A) and sin(A) related to the motion of a rotating dot around a circle?

B: What do the sine and cosine functions look like when they are plotted as a function of time?

C: How are the functions cos(A) and sin(A) related to each other?

D: What happens to the shape of the signal in time when you increase the fundamental frequency of a sine or cosine signal? How does this action change the sound of these signals?

Summary

This lab connects circles, trigonometric functions, and sound waves. Sines and cosines were the first waveforms used in synthesizing music electronically, that is, without having some physical device vibrating back and forth. And, if you were good at moving the slider back and forth in this lab, you could make real music with it. There are better ways to control the frequency content of musical signals, though, as we'll soon learn.

2.3 Listening to Sines and Cosines

Lab Objectives

In previous labs, we have had a chance to see how sines and cosines are related to each other and to musical sounds. In this lab, we get a chance to really put sine and cosine signal generators through their paces. We can set their amplitudes. We can pick their frequencies. We can even adjust their phases. Which of these really matter when it comes to music? Discover for yourself in this lab.

Textbook Reading

*   This lab appears on page **68** of the *Engineering Our Digital Future* textbook.
*   Prerequisite textbook reading before completing this lab: pp. **44-68.**
*   Plotting Examples of Equation (2.12) for different amplitude values for A, frequency values for f, and time shifts d is good preparation for doing this lab.

Additional Materials

*   Graphing Calculator

Engineering Designs and Resources

Worksheet used in this lab:

*   **L02-03-01 Listening to Sines and Cosines.Lst**: Allows you to make sine and cosine functions, plot them, and hear them as well.

Worksheet Description

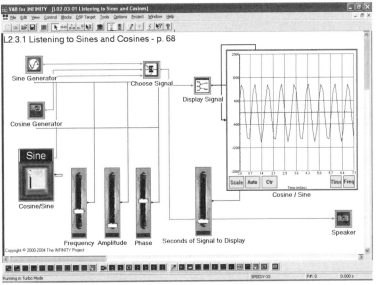

Figure 2.5 Listening to Sines and Cosines

Open the worksheet *L02-03-01 Listening to Sines and Cosines.Lst.*

The worksheet contains two blocks which generate the sine and cosine signals, respectively. When the worksheet is running, both signals are being generated, but you can only see one of them at any one time.

The button switch on the left of the worksheet controls which signal (sine or cosine) is being displayed and heard.

There are three sliders grouped on the lower left of the worksheet. These sliders control the amplitude, frequency, and phase of the cosine or sine signal that is selected.

The plot on the right shows the signal itself. Another slider next to the plot controls how much of the signal is displayed. Only the most-recent portion of the signal is displayed.

Music is made up of sounds that change in amplitude, frequency, and phase. The fundamental frequency of a musical sound is determined by its period, and we hear changes in frequency as changes in pitch.

The block on the bottom right of the worksheet is the Speaker block; it allows us to hear what the cosine or sine signals sound like.

Laboratory Procedures

Steps	Instructions
Step 1:	Start the worksheet.
Step 2:	You should see a plot appear in the display on the right of the worksheet. You should also hear a sound.
	Q1: What is the display showing you?
	Q2: What does the signal sound like?
Step 3:	The slider nearest the display controls how much of the signal that is shown. Move this slider up and down and watch the display.
	Q3: Describe what you see in the display.

Steps	Instructions (Continued)
Step 4:	The parameters of the type of function that you see and hear is controlled by the sliders on the bottom of the worksheet as well as by the button switch. Click on the button switch until it shows a sine function. The button controls whether the function is a sine or cosine. Then, while watching the display, click on the button again.
	Q4: Why does it appear that the signal in the display does not change?
Step 5:	Listen to the sound carefully as you switch back and forth between the sine and cosine signals.
	Q5: Can you hear the any difference between these two signals?
Step 6:	We can calculate the fundamental frequency of the cosine or sine signals by measuring the period of the sine or cosine signal in the display and then using the relationship f = 1/T to calculate the frequency. But, since we have a slider that controls exactly how much of the signal is displayed, you can use the slider to measure the period. Set the Frequency slider to 500, the amplitude slider to 1000, and the Phase slider to 3.14. Next, adjust the Amount of Signal to Display slider so that a single cycle of the sine or cosine function is showing in the display.
	Q6: What is the period of the signal?
Step 7:	Using this period value, calculate the fundamental frequency of the signal.

Steps	Instructions (Continued)
	Q7: What is the fundamental frequency of the signal?
	Q8: How does the value of the fundamental frequency compare to the "dialed in" value in the Frequency slider?
Step 8:	Change the amplitude of the signal from 1,000 to 2,000.
	Q9: How does the signal in the display change as you adjust the amplitude?
	Q10: What about the sound of the signal?
Step 9:	Now, change the amplitude of the signal from 1,000 to 200. Change the phase of the signal from 3.14 radian to 0 radian.
	Q11: What has happened to the volume of the signal?
	Q12: Does the signal in the display change?

Steps	Instructions (Continued)
	Q13: What about the sound of the signal? How did it change?
Step 10:	Using a graphing calculator, plot the function {A cos(2 π f t)} from t = 0 to 0.003 seconds for the following values: A = 2,500, f = 1,200. Then, while the worksheet is running, select the **Cosine** function, change the amplitude of it to 2500, and change the frequency of it to 1,200 Hz. Finally, change the Amount of Signal to Display to 3 msec. Compare the signal in the display with what you see in your graphing calculator.
	Q14: How similar are they?
	Q15: What is different between the two plots?
Step 11:	The display on the right has the capability of calculating the spectrum of the signal provided to it. The spectrums show you the sinusoids that make up a signal. The position of the peaks of the spectrum are at the frequencies of these sinusoids. Click on the **Freq** button to activate this display mode. Describe what you see in the display. The position of the peak of the function you now see should correspond to the frequency of the sinusoid to which you are listening. Change the Frequency slider to a value of 400.
	Q16: What happened to the position of the peak in the display?
Step 12:	Now change the amplitude from a value of 2,500 to a value of 1,000.

Steps	Instructions (Continued)
	Q17: What has happened to the spectrum within the display? You may need to use the "Auto" scaling feature to get the spectrum to fit within the display. Click on the **Auto** button to change its scale.
Step 13:	While still looking at the spectrum, change the signal from a sine to a cosine, or, if you are using a cosine function, change it to a sine function.
	Q18: What happened to the plot in the display?
	Q19: What can you say about the spectrum of the cosine and sine signals of the same frequency?
	Q20: If you wanted to create a simple melody, like "Mary Had A Little Lamb," which of the sliders would you have to change with time: Frequency, Amplitude, or Phase?
Step 14:	Stop the worksheet. Close the worksheet.

Overview Questions

A: What are the names of the three parameters that define a sinusoidal signal?

B: Of these three parameters, which one changes the pitch of the sinusoid the most? Which one changes the sound of the sinusoid the least?

C: How can the spectrum be used to figure out the fundamental frequency of a sinusoid?

D: How is amplitude related to the loudness of a sinusoidal signal?

Summary

This lab allowed us to hear and adjust the shapes of the sine and cosine signals virtually in the VAB environment. By now, you can probably guess that frequency is very important in making musical sounds, and phase plays a much smaller role. In later labs, we'll let the computer control frequencies of sinusoids to make musical sounds.

2.4 Measuring a Tuning Fork

Lab Objectives

A tuning fork is a metal device that vibrates in such a way as to produce a sinusoidal sound of a precise frequency. Musicians, especially acoustic guitar players, use tuning forks to tune their instruments—hence their name. In this lab, we going to investigate the sound that a tuning fork makes; and we'll use a cosine generator to create a nearly-identical copy of the sound. This is an example of modeling: creating an artificial and mathematical version of a real-world phenomenon.

Textbook Reading

* This lab appears on page **69** of the *Engineering Our Digital Future* textbook.
* Prerequisite textbook reading before completing this lab: pp. **65-69.**

Additional Materials

* Tuning Fork or the ability to whistle.

Engineering Designs and Resources

Worksheet used in this lab:

* **L02-04-01 Measuring a Tuning Fork.Lst**: This lab allows you to recreate a tuning fork sound.

Worksheet Description

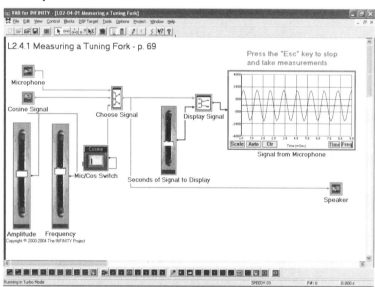

Figure 2.6 Measuring a Tuning Fork

Open the worksheet *L02-04-01 Measuring a Tuning Fork.Lst.*

This worksheet is very similar to the one used in Lab 2.3, "Listening to Sines and Cosines." In fact, the main difference between it and the previous one is the Microphone block, which allows you to listen to and display the microphone signal as heard by the DSP board.

It has a block to generate a cosine signal. The two sliders on the bottom right allow you to change the amplitude and frequency of this cosine. The button just the right of these sliders allows you to choose which signal to see and hear on the worksheet - the microphone signal or the cosine signal.

The slider next to the plots controls how much of the microphone and cosine signals are displayed.

The block on the bottom right of the worksheet is the Speaker block; it allows us to hear what the cosine or sine signals sound like.

Laboratory Procedures

Steps	Instructions
Step 1:	Start the worksheet.
Step 2:	Click on the **Mic/Cos Switch**. You should start hearing a soft tone coming out of the loudspeaker.
Step 3:	Using the frequency sliders, change the frequency of the cosine signal to 500 Hz, and measure the period of the cosine signal on the lower display on the right.
	Q1: What is the period of the cosine signal?
Step 4:	Now, change the amplitude of the signal so that it is equal to 1,000 and hit the **Freq** button on the lower display.
	Q2: At what frequency is the peak of the cosine signal located?
	Q3: Does this make sense?
Step 5:	The Mic/Cos switch can be used to switch between the cosine signal and the micro-phone signal. Set this switch to "Microphone" and talk into the microphone on the DSP board.

Steps	Instructions (Continued)
	Q4: What is happening in the upper display on the right?
Step 6:	Change the scale of the display by using the "Auto", "Scale", and "Ctr" controls so that your speech signal can fit in the display. These controls are buttons that can be pushed and set. Then, whistle into the microphone.
	Q5: What do you see in the display?
Step 7:	While whistling into the microphone, halt the worksheet.
	Q6: What is the period of your whistle?
Step 8:	Next, using the **Freq** button, measure the fundamental frequency of your whistle.
	Q7: What is the value of the fundamental frequency?
	Q8: How is it related to the period of your whistle?

Steps	Instructions (Continued)
Step 9:	Obtain a tuning fork from your teacher. You can whistle instead if you don't have a tuning fork. Tuning forks make sound only after you cause the two bars or tines of the fork to vibrate. The best way to start them vibrating is to hit them on a rubber-type surface, like the bottom of your shoe. Then, hold the bottom end of the tuning fork against a hard surface like a tabletop. You should hear a nice loud tone. Practice getting sound from your tuning fork a few times so that you are ready to take measurements with the DSP board.
Step 10:	Now, we're ready to recreate the sound of your tuning fork or whistle. Set the "Mic/Cos Switch" to Microphone. Make the tuning fork or whistle sound. Hold the microphone near the surface where the fork meets the tabletop. You should see a nice-looking signal in the display.
	Q9: What does this signal look like?
Step 11:	By stopping the worksheet, you can freeze the display to take measurements.
	Q10: What is the period of the tuning fork or whistle signal?
	Q11: What is the frequency of the tuning fork or whistle signal?
	Q12: What is the amplitude of the tuning fork or whistle signal?

Steps	Instructions (Continued)
Step 12:	The cosine generator can be used to make a pretty accurate copy of the tuning fork signal or whistle. From the frequency and amplitude values that you calculated, change the frequency and amplitude of the cosine to match that of the sound. Now, set the switch to the "cosine" setting. You should start hearing the cosine signal from the loudspeaker. Make a sound with the tuning fork or whistle again.
	Q13: How similar are the tuning fork or whistle sound and the cosine signal sound?
	Q14: How similar are their signals in the display?
Step 13:	All tuning forks are designed to create a sound of a note on a piano. The figure below shows the frequencies of some common notes. (See also Figure 2.17, p. 53 in textbook). 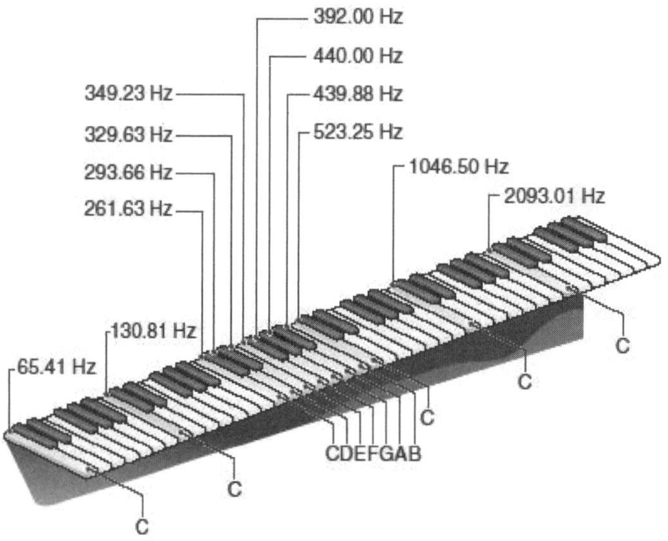
	Q15: What note is your tuning fork or whistle "tuned" to?

Steps	Instructions (Continued)
	Q16: Are the loudspeakers making the same sound?
Step 14:	Stop the worksheet. Close the worksheet.

Overview Questions

A: What type of signal does a tuning fork sound most resemble?

B: If you were to describe a tuning fork or whistle by the sound that it makes, which property is more important to describe--amplitude or frequency?

C: Suppose you wanted to make a digital tuning fork or whistle. What would you need?

Summary

A tuning fork is used by musicians to tune their instruments. This lab shows us that tuning forks make signals that are very close to sinusoids. The pureness of the sound is what makes the job of tuning the instrument easy.

2.5 Building the Sinusoidal MIDI Player

Lab Objectives

This lab is where you "put it all together" to make a fully-functioning digital band. You'll be able to choose your own music for it to play using MIDI (Musical Instrument Digital Interface) files. And, unlike other labs, you'll build this one from scratch by connecting blocks together.

Textbook Reading

*   This lab appears on page **76** of the *Engineering Our Digital Future* textbook.
*   Prerequisite textbook reading before completing this lab: pp. **60-75.**

Engineering Designs and Resources

Worksheet used in this lab:

*   **L02-05-01 Building the Sinusoidal MIDI Player Parts.Lst:** Contains all the blocks needed to build the sinusoidal MIDI player.

Worksheet Description

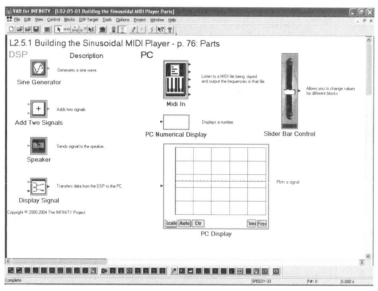

Figure 2.7 Building the Sinusoidal MIDI Player Parts

Open the worksheet *L02-05-01 Building the Sinusoidal MIDI Player Parts.Lst*. You should see something similar to Figure 2.7 above:

Unlike other worksheets that you've used in the past, this one won't run when you hit the green **Go** button. It has a set of blocks that you will use to build up your design. By cutting, pasting, and connecting these blocks together, you will make your own digital band.

The blocks on the left are all DSP-based blocks; this means that they run on the DSP board. The ones on the right are all PC blocks; they run on the PC computer hardware and interact with the DSP board in useful ways. We'll need both types of blocks in our design.

Laboratory Procedures

Steps	Instructions
Building Your Worksheet	
Step 1:	To give us a clean slate to work with, create a new worksheet by hitting the upper-most-left button on the toolbar of VAB.
	This will create a new worksheet that you can save to your own hard disk space on the PC.
	Save this worksheet, perhaps by using your name in the file, such as "Sinusoidal MIDI Player - (My Name).Lst". You can switch back and forth between this new sheet you will build and the parts sheet by selecting them under the window menu.
Step 2:	As in most engineering designs, we'll start from a basic design and move to a more complicated but useful one. Our first design will simply play a single sinusoid out the loudspeaker, where we set the frequency and amplitude of the sinusoid.
	From the Lab 2.5.1 worksheet, copy the Sine Generator block and Speaker block to your new worksheet. Put the Sine Generator block on the left and the Speaker block on the right.
Step 3:	Let's connect these blocks together. To do so, move your mouse cursor to the second grouping of buttons on the VAB toolbar and click on the **Connect Blocks** button (two small squares connected by a single straight horizontal line).
	Now, you are ready to connect blocks together.
Step 4:	Click inside of the Sine Generator block, and after doing so, click on the Speaker block. A connection from the Sine Generator to the Speaker should appear. If it doesn't, repeat the clicking actions in Step 3 and 4. After you are done, go to the button with the arrow on it on the VAB toolbar to get back to **Select** (normal) mode.
Step 5:	Now, you should have a worksheet to make a single sinusoid play sound, but before we run it, let's set the amplitude and frequency of the sinusoid. Double-click on the Sine Generator block to open up its Parameter settings. Out of several entries there, you should see one for Amplitude and one for Frequency.
	Set the Amplitude to 3,000, and set the Frequency to a number that your instructor will give you (s/he can pick a frequency that is different for everyone, so that everyone can play a different note).
Step 6:	Now, you are all ready to make sound. Be careful. It could be very loud if your speakers are set too high, or you might not hear anything because your speakers are off. Make any necessary adjustments and click on **Go** to hear the sound your system makes.
	Stop the system once you've checked that it is working.

Steps	Instructions (Continued)
Step 7:	To see the sinusoidal signal your system is making, you will need the Display Signal, PC Display, and Arbitrary Vertical Slider blocks from your palette. Get these onto your worksheet by cutting and pasting these blocks into your new worksheet, and arrange them as shown in Figure 2.8 below: 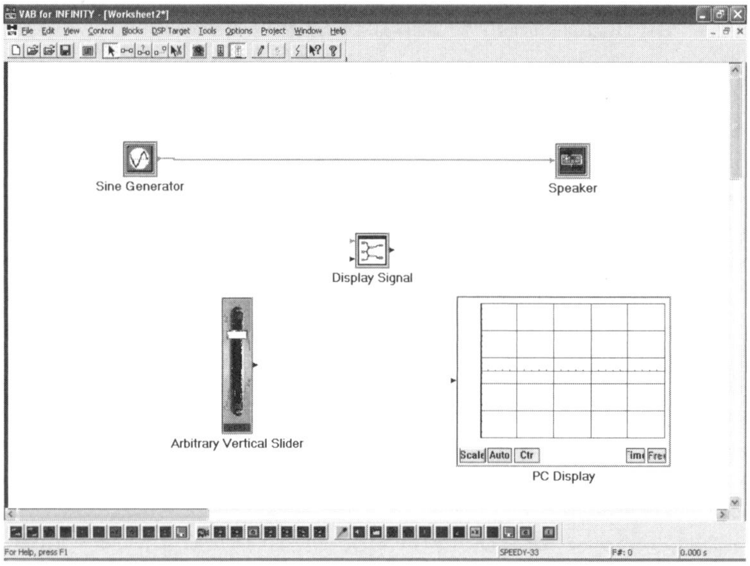 **Figure 2.8** Screen of Step 7
Step 8:	By going through the same steps as in Step 3 and 4, connect the output of your Sine Generator to the top Display Signal input, the Arbitrary Vertical Slider to the bottom Display Signal input, and the Display Signal output to the PC Display. Be sure to click inside of the blocks to get the connections made (your instructor can help you with this part). Be sure to go back to **Select** mode before continuing
Step 9:	You should be all set to run your second worksheet design. Run this worksheet. You should see a sinusoidal signal on the PC Display. Adjust the Arbitrary Vertical Slider until you see a small (say four to seven) periods of the sinusoid.
Step 10:	Sketch the contents of the display, and label your axes carefully. What is the period of your sinusoid?

Steps	Instructions (Continued)
Step 11:	The Sinusoidal MIDI Player uses several sinusoids to make music. Right now, we only have one. So, let's add another Sine Generator to the worksheet. You can copy-and-paste the one that is already there if you want. We'll need the Add Two Signals block, though, from our palette. Once we have both of these blocks on your design worksheet, create the worksheet shown in Figure 2.9 below. 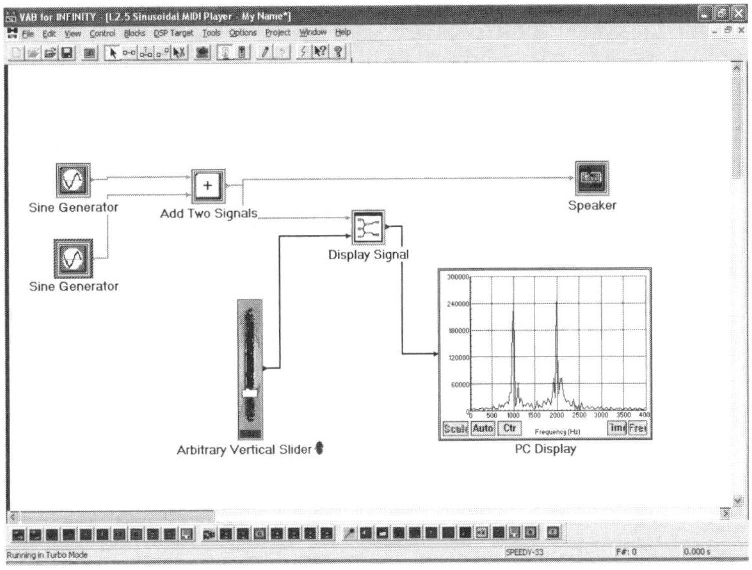

Figure 2.9 Sine Generator added in Step 11

Step 12:	Change the frequency of the second Sine Generator to something different than the first (you can pick the frequency this time, but pick one between 200 Hz and 3 kHz). Once you have the new frequency selected, run the worksheet.
	Q1: Do you hear two sinusoids playing? Sketch the contents of the PC Display after you have stopped the worksheet.

Steps	Instructions (Continued)
Step 13:	Now we are ready to make the five-sinusoid MIDI Player. To do so, we'll need to cut-and-paste enough Sine Generator and Add Two Signal blocks to make five different sine functions and add all of their answers together to get one sound signal. Using similar methods as before, build the worksheet shown in Figure 2.10. You can change the text below each block by clicking on it—it is a good idea to relabel the blocks as you are building so that you can keep track of things. **Figure 2.10** Screen capture of Step 13
Step 14:	Change the frequency of the third, fourth, and fifth Sine Generators so that you have five different frequencies for your Generators. Once you have done this, run the worksheet with the green **Go** button. You should hear all five notes playing at once--not very musical, but it is a proof-of-concept.
Step 15:	To make music, we'll need to have something change the frequencies of the Sine Generators over time. Here is where the MIDI In block comes in. Get this block from your palette, and place it to the left of your five Sine Generator blocks.
Step 16:	The five outputs of the MIDI In block will normally have the frequencies of the sinusoids to be played by the Sine Generators. In order to see the frequency values as actual numbers, we'll use five versions of the PC Numerical Display block from your palette. Create five of these Numerical Displays, and connect each of the outputs of the MIDI In block to its own Numerical Display. At the end of this step, you should have two separate sets of blocks: The MIDI In and Numerical Display blocks connected together; and the Sine Generator, Add, Display Signal, PC Display, and Speaker blocks connected together. You may need to rearrange some blocks in order to see things clearly at this point. Also, renaming the PC Numerical Displays to something meaningful, such as "Frequency 1", "Frequency 2", and so on will help you later.

Steps	Instructions (Continued)
Step 17:	Now, we are ready to have our MIDI In block control the frequencies of our Sine Generator blocks. To do so, we'll use a different connection type: a **Parameter Connect** button on the VAB toolbar. Put VAB into Parameter Connect mode, and connect each MIDI In output to its own Sine Generator block. First, click on the output of the MIDI In block that you want to connect. Then, click in the center of the Sine Generator block you want to control. A dialog box should pop up asking you which variable you want to be controlled. Select **Frequency** and close the dialog box. You should see a green connection appear between the MIDI In block and the Sine Generator block. Do this for all five MIDI In outputs and Sine Generators. The worksheet should look very similar to that in Figure 2.11. **Figure 2.11** Screen capture of Step 17
Running Your Worksheet	
Step 18:	We are ready to play some music. By following directions from your instructor or by referring to **Appendix B**: *Running MIDI VAB Labs* at the back of this manual, get a song playing in MIDI Bar, and then start the VAB worksheet you've just created. You should see frequency values showing up in the PC Numerical Displays; you should hear sound coming from the speaker; and you should see the musical signal being displayed on the PC Display. If you don't see and hear all of these things coming from your worksheet, check with your instructor on your MIDI settings.

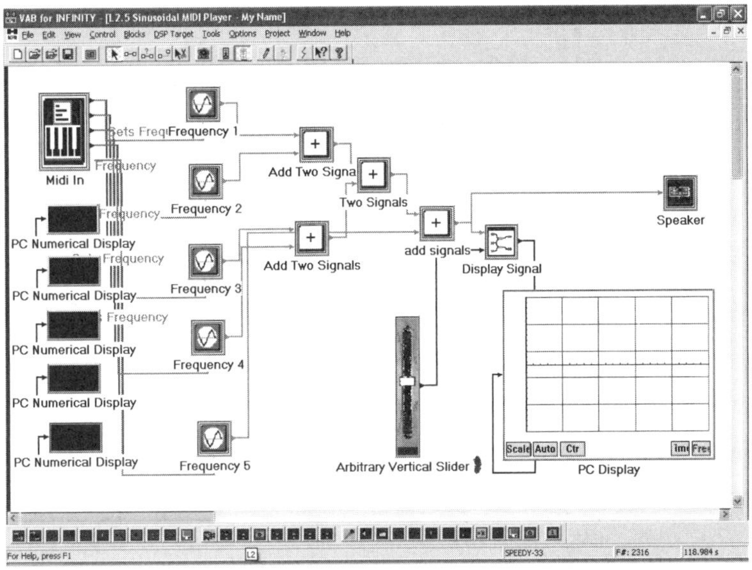

Steps	Instructions (Continued)
Step 19:	All of this music is getting quite loud. We need a way to adjust the volume of all of the Sine Generators in our Sinusoidal MIDI Player. Cut and paste the Slider Bar Control to the worksheet, and relabel this block Volume. Next, double click the slider to change its limits from 0 to 3,000 in 101 steps, and connect its output to the Amplitudes of all five Sine Generators using the Parameter Connect mode. Once you have done this, you will have the worksheet shown in Figure 2.12. 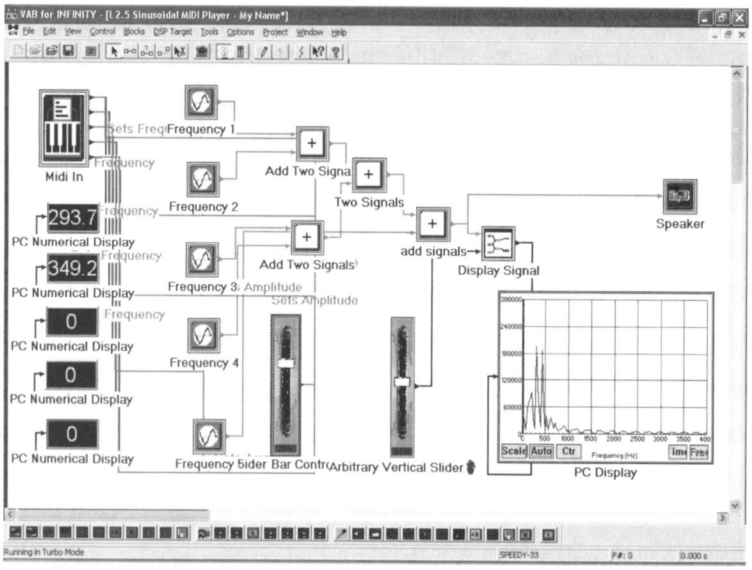 **Figure 2.12** Screen capture of Step 19
	Q2: In looking at the output of the Sinusoidal MIDI Player, does the signal ever look like a sinusoid of a single frequency? How many Frequencies are being detected at the output of the MIDI In block at such times?
Step 20:	The PC Display can display the Spectrum of the sound instead of its time waveform. Simply click on the **Freq** button until it is greyed out.
	Q3: How are the peaks of the spectrum related to the frequencies in the MIDI file?
Step 21:	For fun, go get some MIDI files of your favorite musical group. Your instructor will show you how to do this on the Web. How does your favorite music sound when it is being played with sinusoids?

Steps	Instructions (Continued)
Step 22:	Stop the worksheet.
	Close the worksheet

Overview Questions

A: Which of the following blocks is used to create the basic musical signals in the Sinusoidal MIDI Player? (i) Add Two Signals, (ii) Sine Generator, (iii) MIDI In, (iv) PC Numerical Display, (v) PC Display, (vi) Slider Bar Control.

B: Which of the following blocks is used to read in frequencies to play in the Sinusoidal MIDI Player? (i) Add Two Signals, (ii) Sine Generator, (iii) MIDI In, (iv) PC Numerical Display, (v) PC Display, (vi) Slider Bar Control.

C: Which of the following blocks is used to adjust the volume of the Sinusoidal MIDI Player? (i) Add Two Signals, (ii) Sine Generator, (iii) MIDI In, (iv) PC Numerical Display, (v) PC Display, (vi) Slider Bar Control.

D: How many adders are needed to add five Sine Generators together to make one signal to send to the Speaker block?

E: How could we make the Sinusoidal MIDI Player better? List all of the ways you might want to improve this design.

Summary

This lab showed you that sinusoids can play music. It also was probably your first "build-it" lab. We'll have other "build-it" labs in later chapters as well.

2.6 The Spectogram

Lab Objectives

So far we have looked and listened to cosines. We look at them by plotting their amplitude vs. time. The frequency content or spectrum is a different way of looking at cosines that gives us more insight into what is in the signal. All the cosines we have worked with have a certain frequency. When we talk about the frequency content of a signal, we are asking: What cosines, if added up would give this signal?

Textbook Reading

* This lab appears on page **79** of the *Engineering Our Digital Future* textbook.
* Prerequisite textbook reading before completing this lab: pp. **76-78.**

Engineering Designs and Resources

Worksheets used in this lab:

* **L02-06-01 The Spectrogram File Read.Lst**: Reads a wave file and displays the signal as well as its spectrogram.
* **L02-06-02 The Spectrogram Microphone.Lst**: Adds the chirp signal to the signal from microphone and displays the added signal as well as its spectrogram.

2.6.1 The Spectogram File Read

Worksheet Descriptions

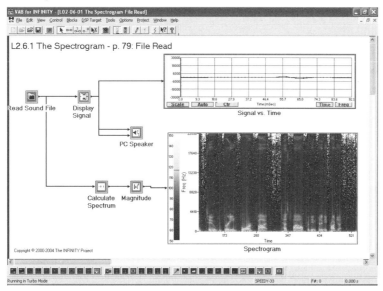

Figure 2.13 Spectrogram File Read

Figure 2.13 shows the Spectrogram File Read worksheet. It's job is simple: Read in a sound file and compute its spectrogram.

Laboratory Procedures

Steps	Instructions
Step 1:	Open up the worksheet *L02-06-01 The Spectrogram File Read.Lst.* Start the worksheet. You should start hearing the contents of the sound file coming out of your PC's computer speakers. You should also see some changes to the "Signal" and "Spectrogram" displays.
	Q1: Is there a relationship between the sounds that you are hearing and the color-coded contents of the "Spectrogram" display? Whenever the voice is silent, what do you see displayed in the "Spectrogram" display?
	Q2: Look at the spectrogram frequency range from 0 to 10000Hz. Yellow and white represent the strongest frequency content. Based on the colors, for this voice file estimate the percentage of the strong frequencies that are below 1,500 Hz. Estimate the percentage of strong frequencies that are between 1,500 Hz and 4,000Hz. Estimate the percentage of strong frequencies that are between 4,000 Hz and 10,000 Hz.
Step 2:	Sketch the bright part of the spectrogram between 0 and 4,000 Hz for the time when the voice is saying "one". Make similar sketches for the time when the voice is saying "two", "three", "four" and "five".
Step 3:	Stop the worksheet. Close the worksheet.

2.6.2 The Spectogram Microphone

Worksheet Description

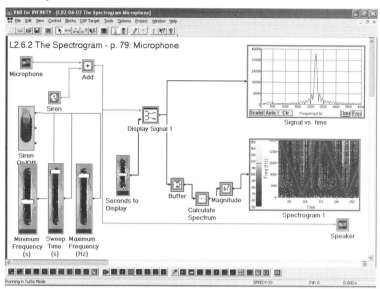

Figure 2.14 Spectrogram Microphone

Open the worksheet *L02-06-02 The Spectogram Microphone.Lst.*

Figure 2.14 above shows the Spectrogram Microphone worksheet. This worksheet creates a siren sound and mixes it with the signal coming from the microphone before displaying the signals two ways: (1) As a plot of either the signal's time or frequency content over a short time in the "Signal" display and (2) as a color-coded image in the "Spectrogram" display. We've seen displays of signals as functions before, but the Spectrogram gives us another way to visualize a signal.

Laboratory Procedures

Steps	Instructions
Step 1:	Start the worksheet.
Step 2:	Adjust the Minimum Frequency and Maximum Frequency sliders until they are at the same value (e.g. 2000 Hz). Then, turn on the "Siren" by flipping the "Siren On/Off" switch to its Up position.
	Q1: What do you hear out of the loudspeakers?

Steps	Instructions (Continued)
	Q2: What do you see in the "Spectrogram" display?
Step 3:	On the Signal Plot click the **Auto** button so that **Scale** is gray. This automatically adjusts the scale so you can see the whole plot. Click the **Freq** button so that it is grayed out.
	Q3: What do you see on the "Signal" plot? How does this plot relate to the frequency content of the signal being made?
Step 4:	Adjust the "Sweep Time" slider until it shows a value of 3 seconds, and then slowly adjust the "Maximum Frequency" to be 3 kHz.
	Q4: Describe what you see in the "Spectrogram" display.
	Q5: What do you hear coming from the loudspeakers? How does this sound relate to the "Spectrogram" display and the changing "Signal" display?
Step 5:	Set the "Siren On/Off" switch to its off position, and talk into the microphone.
	Q6: Describe what you see in the "spectrogram" display.
Step 6:	Stop the worksheet. Close the worksheet

Overview Questions

A: What is different about the spectrum of the signal and its spectrogram? Which one typically measures the frequency content of a signal over a very short time period? Which one displays the frequency content of the signal over a long time period?

B: What will a signal be and sound like if the spectrogram consists of just one horizontal line?

C: What does the spectrogram of a signal such as a human voice or a bird chirp look like? What functions do each of the lines on the spectrogram plot represent?

D: Plot the spectrogram of a single sinusoid with a frequency of 300 Hz.

E: Plot the spectrogram of the following signal:
$s(t) = 3 \cos (2 \pi 100 t) + 1.3 \cos(2 \pi 50 t) - 10 \cos (2 \pi 200 t).$

Summary

The spectrogram is a way to see the frequency content of a signal over a long time, such as several seconds. It is a way to get a picture from a sound that tells us how a sound changes. Spectrograms of simple signals, such as sinusoids, look correspondingly simple, whereas the spectrogram of a complicated sound, such as your voice, can look very intricate. Try thinking of some ways to use the spectrogram in a real-world application.

2.7 Sketch Wave with MIDI

Lab Objectives

The Sinusoidal MIDI Player that you created and/or used in Lab 2.5 made music by adding sinusoids together. This lab takes that idea one step further by allowing you to specify the shape of the periodic signal that makes each note.

Textbook Reading

- This lab appears on page **85** of the *Engineering Our Digital Future* textbook.
- Prerequisite textbook reading before completing this lab: pp. **81-85.**

Engineering Designs and Resources

Worksheet used in this lab:

- **L02-07-01 Sketch Wave with MIDI.Lst**: Allows you to draw a waveform of your choice and plays a MIDI file using this waveform.

Worksheet Description

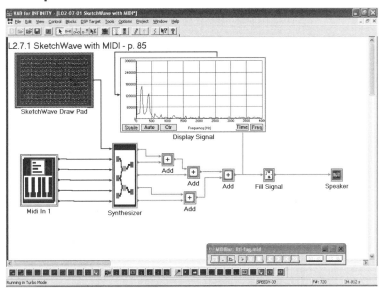

Figure 2.15 Sketchwave with MIDI

This worksheet is quite similar to the one you made in Lab 2.5: Building the Sinusoidal MIDI Player. The main difference is the Synthesizer block, which replaces the five Sine Generator blocks of the Sinusoidal MIDI Player. This block has six inputs: five for the five frequencies that the MIDI In block detects within the MIDI file at any one time, and one for the shape of the periodic function to play. The sketchpad in the upper left-hand corner of the worksheet is where you can sketch the waveform for the synthesizer.

Laboratory Procedures

Steps	Instructions
Step 1:	By following instructions in *A Beginner's Guide to Installing MIDI Tools and Running MIDI-VAB Labs* document, get a MIDI file ready to play using MIDI BAR and the parameter settings of the MIDI In block.
Step 2:	Load a single period of a sinusoidal function in the sketch pad.
Step 3:	Start MIDI BAR running, and then hit the green **Go** button in the VAB software. You should start hearing music playing out the loudspeakers of the DSP Kit.
	Q1: Does the music that you are playing sound similar to that of your Sinusoidal MIDI Player?
Step 4:	While the system is making music, re-sketch the periodic function p(t) in the sketch-pad.
	Q2: What is happening to the sound? Do changes in your sketch pad cause the music sound to change? What changes about the sound--the notes or the type of instrument being played?
Step 5:	Repeat Steps 2 and 3 for the triangle wave and square wave.
	Q3: How different do these sound from the sinusoid?
Step 6:	Try to sketch in the sketch pad one period of the saxophone sound in Figure 2.39 on page 87 of the class text.

Steps	Instructions (Continued)
	Q4: Do you hear music being played by saxophones? If not, what instruments have you created?
Step 7:	Stop the worksheet. Close the worksheet.

Overview Questions

A: How can the shape of a musical instrument sound be used to make multiple versions of that sound to play interesting music?

B: What block determines the period of the sound signals that the Synthesizer block creates: The sketchpad block or the MIDI In block?

C: Suppose you wanted to create a worksheet that would play both the sound of a sinusoid and the sound of a saxophone for every note of a MIDI file. How would you do it? Sketch out the block diagram for this design.

Summary

This is only one of many ways to synthesize sounds. Try and think of your own ways of improving the basic MIDI player.

2.8 Sketch Waves with Envelope Functions

Lab Objectives

This lab gives you a simple way to synthesize sounds using the equation $s(t) = e(t) \times p(t)$, where $e(t)$ and $p(t)$ are functions you sketch out with a mouse on "sketchpads" within a computer display. With a little effort, it is possible to make very realistic musical instrument sounds in this lab just by studying their coarse and fine signal shapes.

Textbook Reading

* This lab appears on page **92** of the *Engineering Our Digital Future* textbook.
* Prerequisite textbook reading before completing this lab: pp. **82-92.**

Engineering Designs and Resources

Worksheet used in this lab:

* **L02-08-01 Sketch Wave with Envelope Functions.Lst**: Creates a sound signal from sketches of its periodic and envelope functions.

Worksheet Description

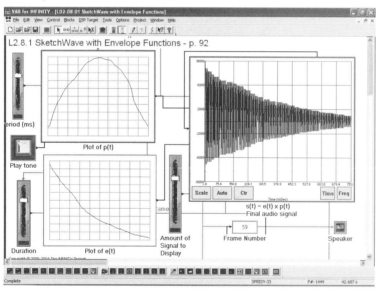

Figure 2.16 Sketch Wave with Envelope Functions

Open the worksheet *L02-08-01 Sketch Wave with Envelope Functions.Lst.*

Figure 2.16 above shows the worksheet used for this lab. The visible worksheet area is divided into roughly half, with the left side containing the user controls and the right side containing the worksheet's display and Speaker icon.

The two grid-like displays on the left of the worksheet are actually "sketchpads" similar to that used in Lab 2.7.1. Here is where you can draw functions for the periodic function $p(t)$ and the envelope function $e(t)$ of the sound you want to create with your computer mouse.

The two sliders "Period (ms)" and "Duration" control the period of the sound that you are making (or the time scale of the $p(t)$ sketch) and the envelope duration (or the time scale of the $e(t)$ sketch).

The **Play Tone** button allows you to hear the result of your sketch--just click on this button and the worksheet will make your s(t) sound.

Laboratory Procedures

Steps	Instructions
Step 1:	Start the worksheet.
Step 2:	Load the same sinusoidal shape a function in the p(t) sketchpad that corresponds to one period of a sinusoidal signal. Then, click **Play Tone**.
	Q1: What do you hear out the loudspeakers? What do you see in the display on the right? You may need to adjust the Amount of Signal to Display slider to "zoom in" on the sound. What have you created?
Step 3:	While keeping the same sinusoidal shape in the p(t) sketchpad, draw a function in the e(t) sketchpad that is approximately exponential from the upper left corner to the lower right corner. Then, click **Play Tone**.
	Q2: Describe in words what is happening in the display on the right and what you hear as well. How is this signal different from the one you made previously?
Step 4:	Adjust the "Duration" slider until the note that you play lasts about one second. Adjust the "Period" slider to 5 ms.

Steps	Instructions (Continued)
	Q3: What does your synthesized sound sound like? Compare the e(t) and p(t) functions you've created to those in Figure 2.15 on page 51 of the class text. **Figure 2.17** Figure on p. 51 of text. Do they look similar? What instrument do you expect this synthesized sound to most resemble?

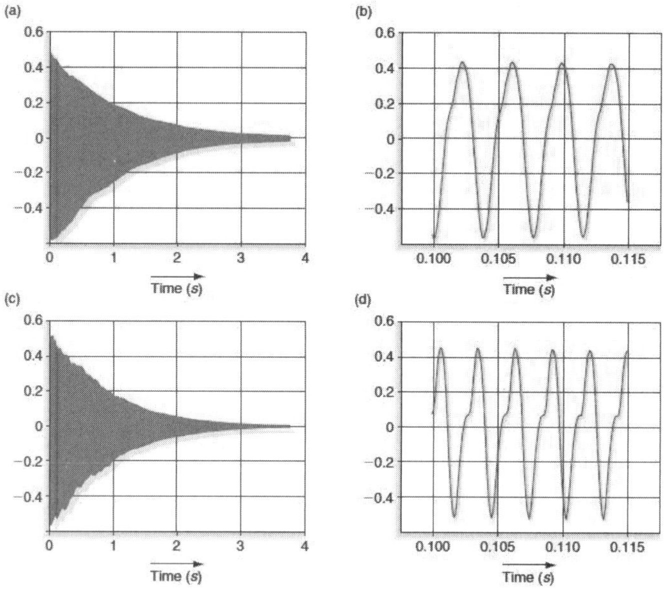

Steps	Instructions (Continued)
Step 5:	Draw a function that is similar to one period of the waveform in Figure 2.39(a) in the p(t) sketchpad, and draw a function that looks like an inverted "U" in the e(t) sketchpad. Adjust the Period slider to 2 ms, and play this sound. 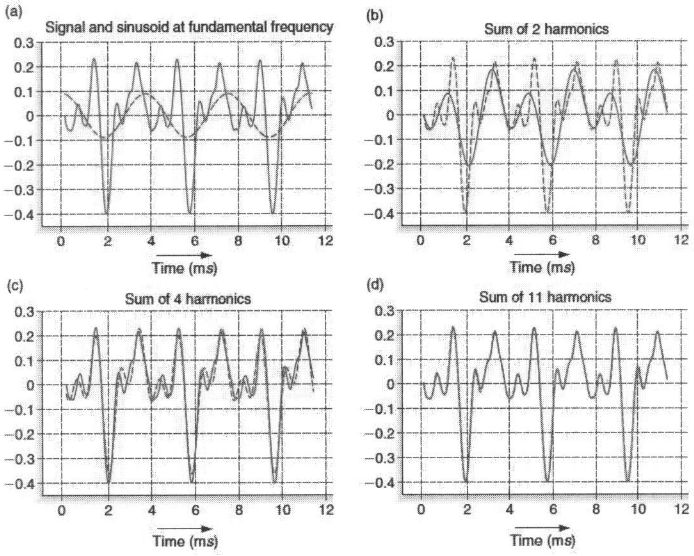 **Figure 2.18** Figure 2.39 of text
	Q4: What does this sound most resemble?
Step 6:	Now, move the "Period" slider to 4 ms and play the sound.
	Q5: What just happened to the sound?

Steps	Instructions (Continued)
Step 7:	Here's your opportunity to be a sound designer.
	Try drawing different waveforms for both p(t) and e(t) in their respective sketchpads. A good rule of thumb to follow when sketching p(t) is: the function should begin and end at the same value on the y-axis; also, the function does not need to repeat to sound musical. As for example e(t) functions, try the various functions in Figure 2.40 on page 89 of the text. Or, have a friend who plays an instrument bring it to class one day. You can use the "Plots of Speech" laboratory worksheet to capture the sound of the musical instrument and display its periodic function p(t) and its envelope function e(t). Using a printout or sketch of the "Plots of Speech" displays, draw your own p(t) and e(t) functions into the "SketchWave with Envelope Functions" worksheet and play the sound. How good is your synthesizer?
Step 8:C	Stop the worksheet.
	Close the worksheet.

Overview Questions

A: When synthesizing sounds using the s(t) = e(t) x p(t) method, which value determines the frequency of a sound, the period of its periodic function p(t) or the duration of its envelope function e(t)?

B: Describe the periodic portion of a piano note sound resemble?

C: Describe the envelope of a piano note sound?

D: Practically speaking, does changing the envelope function of a sound change the fundamental frequency of the sound?

Summary

This lab showed you that interesting musical sounds can be made by simply multiplying two functions together.

2.9 Echo Generator

Lab Objectives

Echo is what you experience when sound comes back to you after bouncing off of a distant wall, canyon, or other hard surface. You also sometimes hear echos caused by an equipment malfunction when making long distance telephone calls. This lab shows you how to simulate echo electronically using the Infinity Technology Kit.

Textbook Reading

*   This lab appears on page **92** of the *Engineering Our Digital Future* textbook.
*   Prerequisite textbook reading before completing this lab: pp. **92-93.**

Engineering Designs and Resources

Worksheet used in this lab:

*   **L02-09-01 Echo Generator.Lst**: Adds echo to any microphone signal.

Worksheet Description

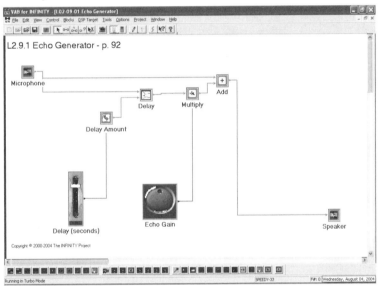

Figure 2.19 Echo Generator

Open the worksheet *L02-09-01 Echo Generator.Lst*. You should see the worksheet in Figure 2.19 above. Let's study this worksheet to see how it works.

On the left is a Microphone block. It senses sounds coming from the Infinity Technology Kit's left microphone. This signal is added to a copy of this signal that has been delayed by a time delay (specified by the slider on the lower left) and multiplied by a number (as specified by the knob in the lower middle portion of the worksheet). The sum of these two signals is sent to the speaker.

Laboratory Procedures

Steps	Instructions
Step 1:	Start the worksheet, taking care to move the microphone away from the loud-speaker of the Kit to avoid feedback. Adjust the Delay slider to its maximum value of 0.5 seconds and the Echo Gain to one.
	Q1: Describe what you hear. What does an echo sound like?
Step 2:	Adjust the Echo Gain to different values between zero and two.
	Q2: How does the sound change?
Step 3:	Set the Echo Gain to one, and change the delay to a value of 0.03 seconds.
	Q3: Can you hear the delayed signal? What has happened?
Step 4:	Stop the worksheet. Close the worksheet.

Overview Questions

A: What is the smallest amount of delay for which you can tell there is a distinct echo for an echo gain of one?

B: Sound travels at approximately 332 meters per second. How far away must you be from a wall to have an echo delay of 0.5 seconds from sound reflecting off of that wall?

Summary

Echoes are caused by reflected sounds off of far away surfaces. Using sound, we can tell how far away something is, too.

2.10 Sound Effects

Lab Objectives

Echo is but one effect used by professional musicians to make great musical sounds. This collection of worksheets allows you to hear some of these other effects and explore their behaviors with your own voice.

Textbook Reading

* This lab appears on page **93** of the *Engineering Our Digital Future* textbook.
* Prerequisite textbook reading before completing this lab: pp. **92-93.**

Engineering Designs and Resources

Worksheets used in this lab:

* **L02-10-01 Sound Effects Reverberation.Lst**: Adds a repetitive echo to a sound.

Worksheet Description

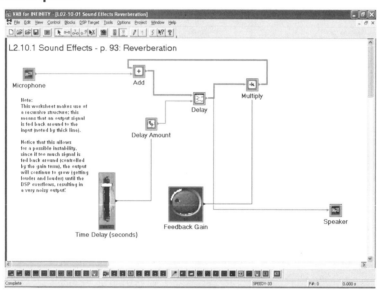

Figure 2.20 Sound Effects - Echo

Open the worksheet *L02-10-01 Sound Effects Reverberation.Lst* and examine its contents.

Notice that this worksheet is quite similar to the Echo worksheet in the last lab, except that the output of the adder is delayed, not its input, before the sum is added to the microphone input signal. By following the path of the adder output signal, you will see that it makes a loop - hence, we expect to hear echoes that repeat over and over.

Laboratory Procedures

Steps	Instructions
Step 1:	Set the Delay amount to 0.5 seconds, and set the Feedback Gain to 0.7.
Step 2:	Start the worksheet.

Steps	Instructions (Continued)
	Q1: Describe what you hear. How is this different from the Echo Generator's sound?
Step 3:	Adjust the delay to a value of 0.05 seconds.
	Q2: Describe what you hear. Does it sound different from the Echo Generator's sound with similar Delay and Gain values?
Step 4:	Set the Time Delay to 0.5 seconds, and set the Feedback gain to one.
	Q3: What happens? Do you expect this? Describe what is going on (after turning off the worksheet)
Step 5:	**Open** the worksheet *L02-10-02 Sound Effects Flanging.Lst*. This worksheet is like the Echo Generator worksheet, except that the fixed Delay has been changed to a variable delay controlled by a Cosine Generator. Run this worksheet and talk into the microphone. Then, blow air over the microphone to simulate a wind sound.
	Q4: What do you hear? How is what you hear related to the Delay value?
Step 6:	**Open** the worksheet *L02-10-03 Sound Effects Stereo Flanging.Lst*. This worksheet also has a variable delay like the Flanging worksheet, but it also has a time-varying balance from left to right speaker. Try running this worksheet, situating yourself between your two speakers.

Steps	Instructions (Continued)
	Q5: What do you hear? How is what you hear related to the Delay value?
	Q6: How is the stereo effect related to surround sound in movies?
Step 7:	Stop the worksheet Close the worksheet

Summary

In this lab you have had the opportunity to experiment further with musical sounds and different effects using your own voice.

Image Quantization

Chapter 3 introduces the basic concepts of digital image representation including capturing, storing, and displaying color digital pictures. These laboratories will help you explore the effects of image representation design choices on the quality and appearance of digital images. Understanding these effects is important when balancing the trade-offs between the quality of a digital image and the amount of storage space it will need.

Infinity Labs

3.1 Image Quantization
3.2 Image Sampling
3.3 Aliasing in Movies
3.4 Color Representation
3.5 Resolution Trade-offs

Introduction

Laborities in chapters 3 and 4 will use VAB blocks for images rather than for sounds. We will still use input blocks, processing blocks, and output blocks as we did in previous laboratories, but these blocks will be designed to work on images. Because images need much more storage space than short intervals of sound signals, we will use the memory and processor of the PC rather than on the DSP board.

Images and video signals are different from sound signals in several important ways. A sound signal is a function of one variable, time, and may continue indefinitely. An image is a function of two variables specifying horizontal and vertical position, and it usually has a limited size or area. A color image is a combination of three different image functions, one for each of the red, green, and blue color components. We listen to a sound signal as it changes over time or we may watch a plot of the time signal to see how it changes. For a single image, we usually display it as a still image and let our eyes explore its content. A video signal is a sequence of still images, so it is a function of three variables, the horizontal and vertical position and also time.

Most VAB blocks for the imaging laboratories are found in the Image Processing Library. The two basic input blocks are an image file reader and the camera. The image file reader block will read a single image from a file stored on the computer in the.bmp format. The camera input supplies a sequence of live images using the camera connected to the PC. The basic output block is an image display block. The display is designed for color images, but it can be used for grayscale images by connecting the same image to the three inputs for red, green and blue. A color image input can be used as three different grayscale image inputs. Usually when a grayscale image is needed, the green image is used. because it is very similar to the grayscale intensity image in most cases.

3.1 Image Quantization

Lab Objectives

In Lab 3.1 we will explore the effects of quantization on image quality. Quantization is one of the two steps needed to represent an image as numbers. One of a limited set of numbers is given to each small square of an image to represent the intensity in that square. We will see that the size of the set of numbers we use will have a big effect on the visual quality of the image and also on the number of bits needed to store the image.

Textbook Reading

*   This lab appears on page **130** of the *Engineering Our Digital Future* textbook.
*   Prerequisite textbook reading before completing this lab: pp. **105-129.**

Engineering Designs and Resources

Worksheets used in this lab:

*   **L03-01-01 Image Quantization Grayscale.Lst**: Demonstrates the effect of image quantization using grayscale images.
*   **L03-01-02 Image Quantization Color.Lst**: Demonstrates the effect of image quantization using color images.

3.1.1 Image Quantization - Gray Scale

Worksheet Description

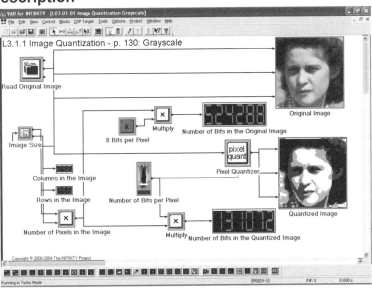

Figure 3.1 Image Quantization Grayscale

Open the worksheet *L03-01-01 Image Quantization Grayscale.Lst.* You should see a screen that looks like Figure 3.1 except that the image displays and numeric displays will be blank.

The worksheet has blocks connected to read an image file from the disk, reduce the number of bits per pixel, and then display the modified image. It also uses several blocks to calculate the total number of bits in the input image and the quantized image.

In the upper left there is a bitmap read block that reads an image file from the disk. This block has no input connections, but it has a parameter to specify the file name of the selected image. This block has three outputs, one each for the red, green, and blue image components. If the selected image is grayscale, then all three components are the same. If the selected image is a color image, the green component is usually a good approximation to the image intensity for a grayscale display.

In the upper right there is an RGB display block to display the image. The block has three inputs, one for each of the three color components. Since we want a grayscale image display, the same image is connected to all three display inputs. This block displays our selected image without any changes, so its inputs are connected directly to the output of the image read block.

The block that is the focus of this laboratory is the pixel quantization block, which can reduce the number of bits per pixel in an image. It has an image input connection, an image output connection, and a parameter connection which controls the number of bits per pixel in the output image. The output of the quantizer is connected to the inputs of a second image display block so the quantized image can be compared to the original image. A slider control block connected to the quantizer block allows you to easily change the number of bits per pixel in the output image while the worksheet is running.

The rest of the blocks are used to calculate and display the total number of bits in the input image and the quantized image. They are included to make it easy to know how much storage is needed.

Note that the "original" image, which is read from the disk, is also quantized. If it were not quantized, we would not be able to store it as an array of numbers and then convert the numbers into an image display. Using this worksheet, we will explore the effects of quantizing the 8-bits-per-pixel input image to create a new image using fewer than 8 bits per pixel.

Laboratory Procedures

Steps	Instructions
Step 1:	Start the worksheet.
Step 2:	You should see the images shown in Figure 3.1 with the slider set to 6 bits / pixel.
	Q1: From the output of the image size block, what is the width and height of the image in pixels?
	Q2: What is the number of bits in the original image? What is the number of bits in the quantized image. What is the ratio of the two?

Steps	Instructions (Continued)
	Q3: How is your answer to Question 2 related to the number of bits per pixel in each?
	Q4: Examine the quantized image carefully. Can you notice any difference between the quantized image and the original image?
Step 3:	Adjust the slider so the number of bits per pixel is 5 and then 4. Observe the changes in the image.
	Q5: At four bits per pixel, what is the ratio of the number of bits in the quantized image to the number of bits in the original image?
	Q6: Examine the different parts of the image at four bits per pixel and indicate whether or not you observe false contouring. When a limited number of gray levels creates the appearance of noticeable edges, that are not related to real edges in objects in the image, this is called false contouring. The water: The pattern of the shirt: The face: The hair: Why do you think you observe false contouring in some parts of the image, but not others?

Steps	Instructions (Continued)
Step 4:	Adjust the slider to 3 bits per pixel. Now the structure of the pattern in the shirt is the same in both images but one looks darker than the other.
	Q7: How do you explain this difference?
Step 5:	Adjust the slider to 1 bit per pixel.
	Q8: How much smaller is this image in terms of bits than the original? How much of the detail from the original image is lost?
Step 6:	Use another image as the original image. With the right mouse button, click the **Read Original Image Block**. You should see the following Bitmap Read Parameters pop-up window: **Bitmap Read Parameters** File Name: girlOnWater512x512.b ∨ Browse Path: C:\Program Files\Hyp ∨ Precision: Integer ▼ OK Cancel Help Click the **Browse** button and then navigate to Program Files\Hyperception\VAB-INF\Images\BWimages and select *cityLights256x256.bmp* from this directory. Click **Open** in the browser's Open window, and then click **OK** in the Bitmap Read Parameters Window.
Step 7:	Adjust the slider to vary the number of bits per pixel from 8 to 1. Observe the results.
	Q9: At low numbers of bits per pixel, how does the detail of this image compare to the detail in the previous image? Why is there a difference?

Steps	Instructions (Continued)
Step 8:	Repeat step 6 and select the image *building256x256.bmp* and then adjust the slider. At first glance, the sky in the original image appears to be the same gray level.
	Q10: How can you demonstrate that there are different gray levels in the sky by adjusting the number of bits per pixel? Using the quantized image, can you estimate the range of eight bit values used for the sky in the original image?
	Q11: Why is the false contouring artifact so much more noticeable in the sky than in the building?
	Q12: Note the total number of bits in the original image.
Step 9:	Repeat step 6 and select the image *building512x512.bmp*.
	Q13: What is the width and height of the image in pixels? How does the number of pixels in the original image here compare to your result in Question 1? Why? How does the image content compare to the image observed in Step 8?

Steps	Instructions (Continued)
	Q14: At 3 bits per pixel what difference do you notice between the original and quantized image in: The cloud? The sky? The buildings? The palm trees? For this image, explain why the difference is more noticeable in some specific parts of the image.
	Q15: What is the value for the number of bits per pixel that makes the number of bits in the quantized image the same as the number of bits in the original image in Question 12? Which image do you think gives you the most information about the scene for the same number of bits in the image?
Step 10:	Try other images and see how the number of bits per pixel is related to image quality.
Step 11:	Stop the worksheet. Close the worksheet.

Optional Quantitative Investigations

Steps	Instructions
Step 1:	Open the worksheet *L-03-01-01 Image Quantization Grayscale.Lst.*
Step 2:	Start the worksheet.

Steps	Instructions (Continued)
Step 3:	Load a test image showing bands of gray levels. Navigate to Program Files\Hyperception\VABINF\Images\BWimages and open *graybands640x480.bmp* in this directory. You should see the following image when the slider is set to 3 bits per pixel. **Figure 3.2** Image Quantization - Grayscale *graybands640x480.bmp*
	Q1: In the quantized image using 3 bits per pixel, which of the eight gray level bands look alike? Why?
	Q2: Reduce the number of bits per pixel to 2. How does the quantized image change? Now how many of the eight gray level bands look alike? Why?
	Q3: Reduce the number of bits per pixel to 1. Explain what happens to the image.
Step 4:	Load another test image. Navigate to Program Files\Hyperception\VAB-INF\Images\BWimages and open *orbs1_256x256.bmp* in this directory. Adjust the number of bits per pixel to intentionally create false contouring.

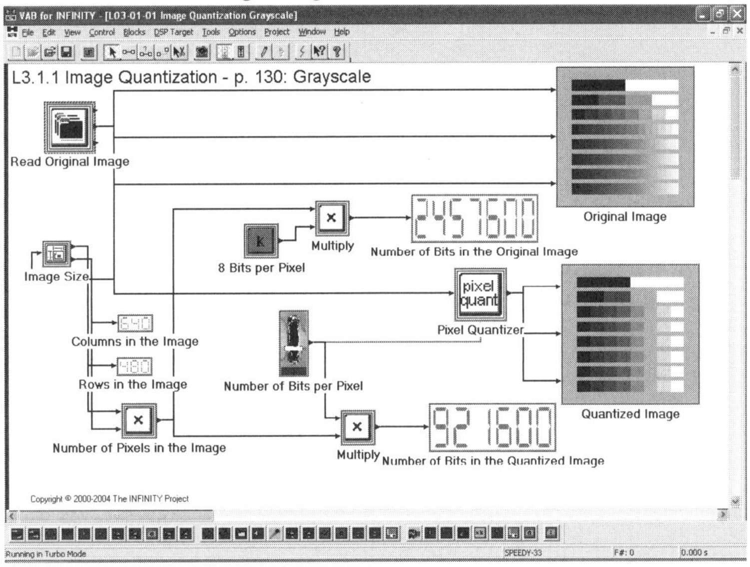

Steps	Instructions (Continued)
	Q4: Do the contours in the image give you more of a sense of the surface detail even though the quantized image looks very different from the original image? What does this tell you about the way the light is reflected from the surfaces in the image?
Step 5:	Load a third test image. Navigate to Program Files\Hyperception\VAB-INF\Images\BWimages and open *orbs2_256x256.bmp* in this directory. Again adjust the number of bits per pixel to intentionally create false contouring.
	Q5: Does this original image look like the original image in Step 4? What do you learn by looking at the quantized image and adjusting the number of bits per pixel?
Step 6:	Stop the worksheet. Close the worksheet.

Optional Further Explorations: How much difference does quantization make?

Steps	Instructions
Step 1:	Open the worksheet *L03-01-01 Image Quantization Grayscale.Lst.*
Step 2:	Load a larger version the image that was used at the beginning of this laboratory. Navigate to Program Files\Hyperception\VABINF\Images\BWimages and select *girlOnWater512x512.bmp.*
Step 3:	By left clicking blocks and dragging them, move the displays of the number of bits per pixel and the multipliers to the left of your worksheet. It is OK for them to overlap other blocks.
Step 4:	Select the output image display by left clicking it. Use **Control C** to copy it and **Control V** to paste a new display into your worksheet. You should now have a blank third display on your worksheet. Right click the label and change it to "Difference Image". We are going to use this display to view the difference between the original image and the quantized image.

Steps	Instructions *(Continued)*
Step 5:	Add two new blocks, **Subtract Image** and **Image Absolute Value,** to the worksheet. From the Blocks drop down menu click **Select Blocks**. Use the Image Processing Library and the Arithmetic Function Group List.
Step 6:	Connect the new blocks: • Connect the Read Original Image green output to the top input of the subtractor block. • Connect the Pixel Quantizer output to the bottom input of the subtractor block. • Connect the subtractor block output to the input of the absolute value block. • Connect the output of the absolute value block to all three inputs of the new image display block. After making all the connections, click the **set-up tool** at the top of the worksheet. (It is the diagonal arrow next to the connection tool.) Your worksheet should look like the figure below. 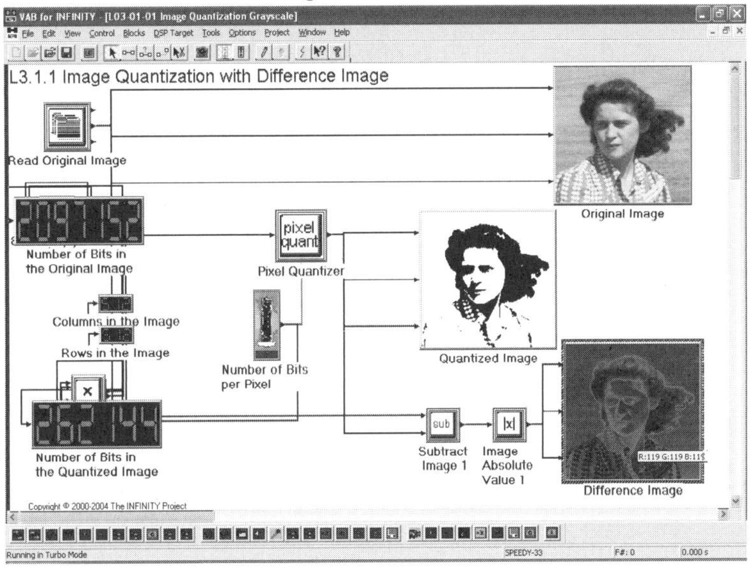 **Figure 3.3** Image Quantization with Difference Image
Step 7:	Start the worksheet. Set the number of bits per pixel to 4.
	Q1: Do you see any difference at all in the Difference Image display? What would the maximum value be for a pixel in that display? By dragging the cursor into the apparently black difference image and left clicking, the actual pixel values can be read.

Steps	Instructions (Continued)
	Q2: Record about ten of these values. Are they consistent with your predictions?
Step 8:	Reduce the number of bits per pixel to 2.
	Q3: What is the maximum value a pixel in the difference image can have?
Step 9:	Using the cursor, find the values of ten pixels that are in what appears to be the brightest parts of the difference image.
	Q4: Are these values consistent with your predictions?
Step 10:	Reduce the number of bits per pixel to 1.
	Q5: How is the appearance of the difference image different from the quantized image.
Step 11:	Stop the worksheet. Close the worksheet.

3.1.2 Image Quantization - Color

Worksheet Description

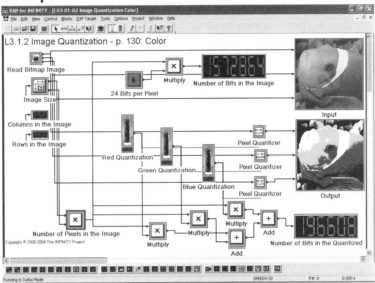

Figure 3.4 Image Quantization Color

Open the worksheet *L03-01-02 Image Quantization Color.Lst.* You should see a screen that looks like Figure 3.4 above except that the image displays and numeric displays will be blank.

This worksheet is very similar to the worksheet for the gray scale image quantization laboratory. Since it uses color images, all three outputs of the image read block are individually connected to the corresponding color inputs for the image display block. The three color components are individually quantized, and the number of bits assigned to each color component can be different.

Laboratory Procedures

Steps	Instructions
Step 1:	Start the worksheet.
Step 2:	You should see the images shown in the figure above with the three sliders each set to 1 bit / pixel.
	Q1: Look at the output of the image size block and note the width and height of the image in pixels.

Steps	Instructions (Continued)
	Q2: How many bits per image are needed for both the original image and the quantized image. What is the ratio of the two?
	How is this related to the number of bits per pixel for each of the three colors of the quantized image?
	What is the minimum number of bits per pixel for a color image?
	Q3: What is the quality of the image both in terms of whether you can tell what the image is and whether it looks realistic?
Step 3:	Slowly increase the number of bits of the blue image component from 1 to 8 and observe the results.
	Q4: When 8 bits/pixel are used for the blue component while 1 bit per pixel is used for red and green, what is the ratio of the total number of bits in the quantized image and the original image?
	Q5: How much did the quality of the image improve by using more bits for blue?
Step 4:	Return the blue component slider to a value of 1, and then observe the results of slowly increasing the value of the green slider from 1 to 8.

Steps	Instructions (Continued)
	Q6: Repeat questions 4 and 5 for this image which now has 8 bits per pixel for the green component and one bit per pixel for the red and blue components.
	Q7: How does the image quality of this image compare to that of the original image?
	Q8: How does the image quality of this image compare to that of the image from Step 3 with 8 bits per pixel used for the blue component?
Step 5:	Return the green component slider to a value of 1, and then observe the results of slowly increasing the value of the red slider from 1 to 8.
	Q9: Repeat questions 4 and 5 for this image.
	Q10: How does the image quality of this image compare to that of the original image?

Steps	Instructions (Continued)
	Q11: How does the image quality of this image compare to that of the images from Step 3 with 8 bits used for the blue component and Step 4 with 8 bits used for the green component.
	Q12: If you could use 8 bits for one color component but only one bit each for the other two, which component would you choose for the 8 bit representation? Is this what you would have expected for an image of a bright orange fish?
Step 6:	Find the best quantized image that you can using 12 bits per pixel instead of the 24 bits per pixel used in the original image. You should be able to get an image that is reasonably close to the original. Adjust the number of bits for the red, green, and blue components so that the sum equals 12.
	Q13: What is the best number of bits to use for each component if the total number of bits per pixel is 12?
Step 7:	Repeat Step 6 using a total of 8 bits per pixel.
	Q14: What is the best number of bits to use for each component if the total number of bits per pixel is 8? Is there a noticeable reduction in image quality when using 8 bits per pixel instead of 12? (There is no one correct answer because individuals will have different ideas about what image is best.)
Step 8:	Load a new color image. Navigate to Program Files\Hyperception\VAB-INF\Images\colorImages and open *building512x512.bmp* in this directory.

Steps	Instructions (Continued)
	Q15: Which color seems to be the most dominant in this image: red, green, or blue?
Step 9:	Explore the effects of adjusting the number of bits per pixel for each color component as you did in the previous image.
	Q16: If you could use 8 bits for one color component but only one bit each for the other two, which component would you choose for the 8-bit representation? Is this what you would have expected for an image where blue is the most noticeable color?
Step 10:	For this image find the best quantized image that you can using only 12 bits per pixel. You should be able to get an image that is reasonably close to the original.
	Q17: What is the best number of bits to use for each component if the total number of bits per pixel is 12?
	Q18: How is this result similar to the answer to **Q13**? How is it different?
Step 11:	Stop the worksheet. Close the worksheet.

Overview Questions

A: Why don't color images need three times as many bits per pixel as grayscale images to have reasonable quality?

Summary

The relationship between the quality of an image and the number of bits per pixel depends to some extent on the content of the image. Images with smooth changes in shades (like a person's face or the blue sky) need more bits to avoid false contouring than areas of an image where colors change rapidly. We can reduce the number of bits per pixel to save storage space, but after a small reduction the image quality is poor.

3.2 Image Sampling

Lab Objectives

In Lab 3.1 we explored the effects of quantization on image quality and on image storage space measured in bits. Now we will discover that selecting the correct number of pixels in an image also has a big impact on image quality and image storage. This is called sampling. Both sampling and quantization are necessary if we want to represent an image as numbers.

Textbook Reading

- This lab appears on page **135** of the *Engineering Our Digital Future* textbook.
- Prerequisite textbook reading before completing this lab: pp. **130-134.**

Engineering Designs and Resources

Worksheets used in this lab:

- **L03-02-01 Image Sampling Grayscale.Lst**: Demonstrates the effect of sampling using grayscale images.
- **L03-02-02 Image Sampling Color.Lst**: Demonstrates the effect of sampling using color images.

3.2.1   Image Sampling - Gray Scale

Worksheet Description

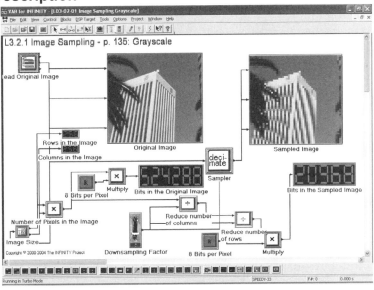

Figure 3.5 Image Sampling - Grayscale

Open the worksheet *L03-02-01 Image Sampling Grayscale.Lst.* You should see a screen that looks like Figure 3.5 except that the image displays and numeric displays will be blank.

It is very similar to the worksheet for the previous laboratory except that the quantization block now has been replaced with a decimate block for this laboratory.

After opening this laboratory you can right click on the decimate block to learn what it does. Then left click the **Help** button and you will see a short description of the block. It has one image input and one image output. The function of this block is to divide the image into n pixel by n pixel squares and then replace each n x n pixel square with a single pixel whose value is the same as one of the pixels in the square. This function is called decimation. Another name for decimation is downsampling.

Using this worksheet, we will explore the effects of reducing the number of pixels in the sampled image.

Laboratory Procedures

Steps	Instructions
Step 1:	Start the worksheet.
Step 2:	You should see the images shown in the Figure 3.5 above but with the slider set to a downsampling factor of 1.
	Q1: When the downsampling factor is 1, why is the sampled image exactly the same as the image read from the disk? (Hint: Remember the description of the decimate block operation from the worksheet description above.)
Step 3:	Reduce the number of pixels in the image by increasing the downsampling factor to 2 and observe the results.
	Q2: How many bits are used in the sampled image? How many pixels are in the sampled image? How is the number of pixels related to the downsampling factor and the number of pixels in the original image?
	Q3: What differences between the sampled image and the original image can be seen in the following areas? -The sky -The top edge of the building -The sides of the building -The palm trees

Steps	Instructions (Continued)
Step 4:	Reduce the number of pixels in the image by increasing the downsampling factor to 3 and observe the results.
	Q4: How many bits are used in the sampled image? How many pixels are in the sampled image? How is the number of pixels related to the downsampling factor and the number of pixels in the original image?
	Q5: What has happened to the appearance of the sides of the building?
Step 5:	Reduce the number of pixels in the image by increasing the downsampling factor to 6 and observe the results.
	Q6: What has happened to the appearance of the sides of the building? Does it still look like a building? Can you still tell that the tree is a palm tree?
Step 6:	Increase the downsampling factor until you can no longer be sure what the image represents and record that number.
Step 7:	Adjust the downsampling factor and fill in the following table:

Downsampling Factor	pixels in sampled image	bits in sampled image
1		
2		
3		
4		
5		
6		

Steps	Instructions (Continued)
	Q7: Plot the number of bits in the sampled image as a function of the downsampling factor. What kind of curve is this? Write a mathematical expression for this curve.
Step 8:	Stand about 12 feet away from the screen and have your lab partner move the slider between 1 and 8 several times.
	Q8: At what values of the downsampling factor do the images look alike at that distance? At what values can you see a difference?
	Q9: Increase your viewing distance from the screen and repeat these observations.
Step 9:	Now we will look at the effect of downsampling on a different image. With the right mouse button, click the **Read Original Image** block. Click the **browse** button and then navigate to Program Files\Hyperception\VAB-INF\Images\BWimages and select *cityLights256x256.bmp* in this directory. Click **Open** in the browser's Open window, and then click **OK** in the Bitmap Read Parameters Window.
Step 10:	Set the downsampling factor to 1 so that the sampled image is the same as the original image. Increase the downsampling factor until you can no longer be sure what the image represents and record that number.

Steps	Instructions (Continued)
	Q10: Why is this number so different from the value you recorded in step 6? Why do you think this particular image was so tolerant to reducing the number of bits per pixel in the previous laboratory but is so sensitive to reducing the number of pixels here?
Step 11:	Read a new image by navigating to Program Files\Hyperception\VAB-INF\Images\BWimages. Select *BABOON256x256.bmp*.
Step 12:	Adjust the downsampling factor until the eye on the left side of the image is a light 3 pixel by 3 pixel square with a black square in the center. Record the value of the downsampling factor.
	Q11: Based on the value you just recorded, estimate the size of the eye in the original image in terms of pixels.
Step 13:	Try other images and see how the number of pixels is related to image quality.
Step 14:	Stop the worksheet. Close the worksheet.

Optional Quantitative Investigations

The unexpected patterns we saw as we changed the downsampling factor for the image of the buildings demonstrates spatial aliasing. When we don't have enough pixels to represent the detail in an image we can get incorrect image patterns. These patterns can also be distracting because a small change in position of the objects being viewed can cause a very large change in the image. This effect is sometimes seen on news broadcasts if a reporter is wearing a striped tie. The pattern on the tie may appear to shift orientation dramatically every time the reporter moves.

We will use a regular diagonally striped pattern with sinusoidal intensity variation to help us understand these effects. Since it is a periodic pattern, we can use the concept of the period of the pattern from Chapter 2. In the case of images, we can measure a period in terms of pixels rather than units of time.

Steps	Instructions
Step 1:	Open the worksheet *L03-02-01 Image Sampling Grayscale.Lst*.
Step 2:	Start the worksheet.

Steps	Instructions (Continued)
Step 3:	Load a test image showing a regular pattern of diagonal stripes. Navigate to Program Files\Hyperception\VABINF\Images\BWimages and open *StripePat1at256x256.bmp* in this directory. You should see Figure 3.6 below when the slider is set to a downsampling factor of 10. **Figure 3.6** Downsampled striped image
Step 4:	Set the downsampling factor to 1 so that the sampled image and the original image are the same. Count the number of cycles of the periodic intensity pattern along the left vertical edge of the image. The easiest way to count the cycles is to find the centers of the light areas. A cycle or a period will be the interval between two of the white centers. (Or, you can find the centers of the black areas and use the same process.) Then count the number of cycles of the periodic intensity pattern along the top horizontal edge of the image in the same way.
	Q1: Estimate the number of pixels in each period or cycle of the pattern on the left vertical edge. Each edge has 256 pixels, so (256 pixels/ number of periods) will be the number of pixels per period. Then estimate the number of pixels in each period or cycle of the pattern on the top horizontal edge in the same way. (Note: The estimate of the number of periods you observe or the number of pixels in a period that you compute might not be an integer.)
Step 5:	Increase the value of the downsampling factor to 2 and then 3 while observing the sampled image display.

Steps	Instructions (Continued)
	Q2: Do the stripes appear to keep the same orientation? Count the number of periods of the intensity pattern along the horizontal and vertical edges. Compute the number of pixels in a period of the horizontal and vertical intensity patterns for a downsampling factor of 2. Repeat for a downsampling factor of 3.
	Q3: What differences do you notice between the original image and the sampled image when the downsampling factor is 3?
Step 6:	Increase the downsampling factor to 6 and observe the sampled image display.
	Q4: Repeat Question 2 for this downsampling factor of 6.
	Q5: Repeat Question 3 for this image.
Step 7:	Switch the value of the downsampling factor between 6 and 7 several times and observe the sampled image.
	Q6: Do the stripes appear to change direction?

Steps	Instructions (Continued)
	Q7: Repeat Question 2 for a downsampling factor of 7.
	Q8: Compare the number of periods in the horizontal direction for downsampling factors of 6 and 7. How does this difference explain your observations from **Q6**?
	Q9: Compare the number of pixels per period in the horizontal direction for downsampling factors of 6 and 7. Can this number ever be smaller than 2? Explain your answer.

Steps	Instructions (Continued)

Step 8: **Measure values:**

Adjust the downsampling factor and count the number of periods on the horizontal and vertical edges. Fill in the following table. Several table entries have been computed already in previous steps. Put a star by the downsampling values that cause the orientation and width of the stripes in the sampled image to be different from the orientation and width of the stripes in the original image. (The first two entries have been done for you.)

Downsampling Factor	pixels along edge of sampled image	Vertical Pattern periods	pixels/ period	Horizontal Pattern periods	pixels/ period
1	256	2.5	102.4	19	13.5
2	128	2.5	51.2	19	6.7
3					
4					
6					
7					
12					
16					

Steps	Instructions (Continued)
Step 9:	**Compute values:** The original image was created using 100 pixels per period in the vertical direction, so the actual number of periods would be 2.56, not 2.5 as measured from the image. (But with the method we are using to count the number of periods, it would be difficult to distinguish 2.5 periods from 2.56 periods.) The period in the horizontal direction was 1/0.075= 13.33 pixels, so the actual number of periods in the horizontal direction is 19.2, not 19 as measured. Using these actual numbers of periods for all downsampling factors, compute the number of pixels/period you would expect to see in the image and enter them in the table below. Compare these results with what you observed in Step 8. What appears to happen in the sampled image display when the computed number of pixels/period falls below 2? Can both these results and the results from Step 8 be correct?

Downsampling Factor	pixels along edge of sampled image	**Vertical Pattern**		**Horizontal Pattern**	
		periods	pixels/ period	periods	pixels/ period
1	256	2.56	100	19.2	13.3
2	128	2.56	50	19.2	6.67
3		2.56		19.2	
4		2.56		19.2	
6		2.56		19.2	
7		2.56		19.2	
12		2.56		19.2	
16		2.56		19.2	

Steps	Instructions (Continued)
Step 10:	Stop the worksheet. Close the worksheet.

Optional Further Explorations: How is the image content related to the number of pixels needed for a reasonable quality image?

Steps	Instructions
Step 1:	Open the worksheet *L03-02-01Image Sampling Grayscale.Lst.*
Step 2:	Add a new block: From the Blocks drop down menu on the top bar, click on **Select Blocks.** In the pop-up window that appears select **Infinity Hot List** for the Library and for the Group List select **Image and Movies**. • In the Function List on the right, click **Camera** and then click the **Add to Work-sheet** button. • Click the **Close** button to close the pop-up window now. You should see a new camera block on your worksheet.
Step 3:	Rewire your worksheet: Click the **deletion tool icon** at the top of the worksheet. (It looks like a pair of scissors.) • Now when you click a block or a wire it will be deleted. • Disconnect the bitmap read block by deleting the wires from it to the decimate block input, the image size block input, and the three inputs to the original image display. Click the **connection tool icon** at the top of the worksheet. (It looks like a horizontal line connecting two points.) • Connect the middle (green) camera output to the three inputs to the original image display. • Connect the middle camera output to the input of the decimate block. • Connect the middle camera output to the input of the image size block.
Step 4:	After making all the connections, click on the **set-up tool** icon at the top of the work-sheet. (It is the diagonal arrow next to the connection tool.) • Right click the **image size** block and check the precision parameter. If it is set to auto, change it to integer. (In auto precision it uses bytes because the image data coming from the camera is in bytes. Byte precision is not appropriate for the image size parameters because it will show all values as the remainder after division by 256.)
Step 5:	Start the worksheet and set the downsampling factor to 4. Place something with a lot of detail such as a cell phone or a wire spiral from a notebook close to the camera and observe both images. You may need to adjust the focus of the camera.

Steps	Instructions (Continued)
	Q1: What happens to the two images as you move the object away from the camera? Explain your observations.
Step 6:	Set the object at a distance from the camera where the original image shows the correct detail but the sampled image shows very noticeable reduction in image quality when the downsampling factor is 4. Record the distance.
	Q2: Estimate the number of pixels used for the object in the original image and the number of pixels used for the object in the sampled image.
Step 7:	Move the object to a distance twice as far from the camera as you had it in Step 8.
	Q3: Estimate the number of pixels used for the object in the original image and the number of pixels used for the object in the sampled image.
	Q4: Move the object four times as far away from the camera as you had it in Step 5 and repeat Question 3.
	Q5: What happens to the number of pixels used for an object as the distance of the object from the camera increase? Why?

Steps	Instructions (Continued)
	Q6: Plot the number of pixels vs. distance from the camera. Plot the square root of the number of pixels vs. distance from the camera.
Step 8:	Move the object back to the distance used in Step 6. Take a good look at the sampled image. Then move the object away from the camera until the number of pixels used for the object in the original image is approximately the same as the number used in the downsampled image before the object was moved. How much further away did you have to move the object?
	Q7: With respect to image detail, compare the image quality of the current original image to the previous sampled image. (The current image will be smaller, but you can make the display a little larger on the screen.) Compare the results of lowering the number of pixels used for an object by adjusting the downsampling factor to the result of lowering the number of pixels used for an object by moving it away from the camera. Consider both the blockiness of the display and the amount of detail you can see.
	Q8: How can the results of Question 7 be explained if the effect of moving an object away from the camera causes a larger surface area to be averaged together to create a pixel value while the effect of downsampling is to select a single pixel from a group of pixels regardless of how different it might be from neighboring pixels?
Step 9:	Explore the effects of different downsampling rates on objects that are moving. For example, put your hand close to the camera and increase the downsampling rate to 8. Observe the image of your hand as you move your fingers.
Step 10:	Stop the worksheet. Close the worksheet.

3.2.2 Image Sampling - Color

Worksheet Description

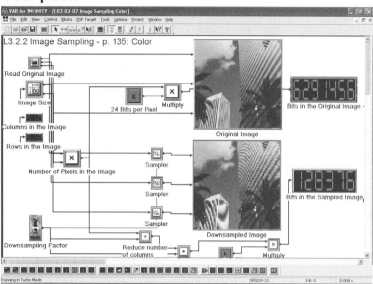

Figure 3.7 Image Sampling - Color

Open the worksheet *L03-02-02 Image Sampling Color.Lst.* You should see a screen that looks like Figure 3.7 above except that the image displays and numeric displays will be blank.

This worksheet is very similar to the worksheet for the grayscale image sampling laboratory. Since it is uses color images, all three outputs of the image read block are individually connected to the corresponding color inputs for the image display block. The three color components are all downsampled by the same factor, so three decimate blocks are controlled by a single slider.

Laboratory Procedures

Steps	Instructions
Step 1:	Start the worksheet.
Step 2:	You should see the images shown in the Figure 3.6 with the downsampling factor set to 1.
Step 3:	Slowly increase the downsampling factor from 1 to 6 and observe the changes in the sampled image.

Steps	Instructions (Continued)
	Q1: With a downsampling factor of 6, examine the sampled image and note differences from the original image in the following areas: -The sky -The top edge of the small building -The sides of the small building -The side of the large building -The palm trees
	Q2: What is the ratio of the number of bits in the sampled image to the number of bits in the original image? Is the loss of image quality worth the savings in storage required if: -You are making a photograph to place in a scrapbook? -You are posting the image on a web page? -You are transmitting the image from Jupiter to Earth at 300 bits per second?
	Q3: Repeat Question 2 for a downsampling factor of 3.
Step 4:	Increase the downsampling factor to 12 and observe the changes in the sampled image.

Steps	Instructions (Continued)
	Q4: With a downsampling factor of 12 can you still tell what the objects in the image are? Examine the sampled image and note differences from the original image in the following areas: -The sky -The top edge of the small building -The sides of the small building -The side of the large building -The palm trees
	Q5: Look more closely at the sky in the original image and the sampled image. In the original image it appears to be the same color everywhere. How does the sampled image show you that there are small color variations in the sky in the original image? Use the cursor to read 5 different RGB values for five different pixels in the sky of the sampled image. Move the cursor to a pixel location and click the left mouse button to see the three RGB values. Record these values.
	Q6: Is this image quality suitable for any application?

Overview Questions

A: Do we ever have enough pixels to perfectly show all the details in an object. Why? If we cannot have enough pixels for a perfect image, how do we decide how many pixels to use?

Summary

The relationship between the quality of an image and the number of pixels used depends on the content of the image. If we do not use enough pixels, the edges of objects will look blocky and the size and orientation of lines may change.

3.3 Aliasing in Movies

Lab Objectives

In this laboratory we will view a video of a bicycle wheel spinning and see several things that we know cannot be true. Parts of the wheel will appear to be still while other parts of the wheel are moving. Different parts of the wheel will appear to be rotating in opposite directions at the same time. These effects, which are called aliasing, occur because the sequence of still frames that make the movie are taken at a rate that is too slow relative to the rotating wheel.

Textbook Reading

- This lab appears on page **137** of the *Engineering Our Digital Future* textbook.

- Prerequisite textbook reading before completing this lab: pp. **135-136**.

Engineering Designs and Resources

This laboratory is a demonstration laboratory and it will not use the VAB worksheets. We will use the media player on your computer to view the video and observe several different examples of aliasing. Initially the video shows a stationary bicycle wheel. The wheel is hand-pedaled to get it rotating at a reasonably fast rate. After the pedaling stops, the rotation rate gradually slows down due to friction. As it slows down we see a variety of aliasing illusions caused by the relative rate of the movie frames and the rate at which the wheel is rotating. We do not have a direct way to measure the speed of rotation of the wheel, but we can tell that it is slowing down by the sound it is making. It may be necessary to view the video several times to notice all the illusions created by aliasing.

Laboratory Procedure

Steps	Instructions
Step 1:	Using Windows Explorer or other navigation tool, go to the Program Files\Hyperception\VABINF\Movies directory.
	Double click *Wheel.mpg* to play the video with the media player utility on your computer.
	If necessary increase the size of the image on the screen using the video size option under View on the toolbar.
Step 2:	Before you start the video, examine the picture of the still wheel. There are three different types of regularly spaced features on the wheel - the reflectors, the spoke connections to the wheel, and the ridges of the tire tread pattern.
	Q1: Count the number of each of these features on the wheel. (You may need to estimate from a count of the features on half of the wheel.) Reflectors Spokes Tire tread ridges

Steps	Instructions (Continued)
Step 3:	Start the video and watch very carefully in the beginning of the video to see which way the wheel is turning. It will turn in this direction throughout the video except for the few seconds at the end. Watch the video and notice that: • At several times the reflectors seem to slow down, then remain stationary, then reverse direction. • At different times the spokes seem to slow down, then remain stationary, then reverse direction. • At some times the reflectors and the spokes seem to be rotating in different directions
Step 4:	View the video several times and use the time reference on the media player to record times when the following events are noticed. You may wish to pause and restart so you can note the times.
	Q2: At what times do the reflectors appear stationary?
	Q3: At what times do the reflectors appear to change directions without stopping?
	Q4: At what times to the spokes appear stationary?
	Q5: At what times do the spokes and the reflectors appears to be rotating in oppo-site directions?

Steps	Instructions (Continued)
	Q6: At what times near the end of the video do the ridges of the tire appear stationary while the spokes and reflectors are moving?
	Q7: What can we say about the relative rate of the movie frames and the revolutions per second of the wheel if the reflectors appear to be stationary?
	Q8: What can we say about the relative rate of the movie frames and the revolutions per second of the wheel if the spokes appear to be stationary?
	Q9: Why do the tire tread ridges appear to be stationary at much lower rotation speeds than the reflectors?
Step 5:	Stop the worksheet. Close the worksheet.

Optional Quantitative Investigations

Steps	Instructions
Step 1:	Do the following computations assuming that the movie was made at 15 frames/sec.
	Q1: Find the 4 smallest values for the rotations per second of the tire that will make the reflectors appear to not be moving.
	Q2: Find the range of values of rotations per second of the tire that make the reflectors appear to rotate in the wrong direction. Restrict your answer to rates slower than the fastest rotating rate you answered in Question 1.
	Q3: Repeat Question 1 for the spokes.
	Q4: Repeat Question 2 for the spokes.
	Q5: Assume that the tire is 1.5m in circumference. At what speeds in m/s would a video of the moving bicycle show the reflectors and wheels rotating in opposite directions?

Overview Questions

A: How can aliasing effects be reduced?

B: How do the limitations of the human visual system affect the selection of frame rates for movies?

Summary

Aliasing effects are seen in movies when the number of frames per second is too low compared to the rate of movement of objects in the movie. This can result in illusions about object motion. It is particularly noticeable in movies of rotating objects.

3.4 Color Representation

Lab Objectives

Color images can be created by mixing three different colors of light. Television screens and computer monitors combine red, green, and blue light to make most of the colors that we normally see. Because there are three components of color, a particular combination of components can be thought of as a specific point in a three dimensional color space or color cube. In the example shown below, each component can have a value of 0, 1, 2, or 3. In this laboratory you can become familiar with how colors are made from different mixtures of the three components by interactively controlling the amount of each component and viewing the result and also by matching colors from a camera image.

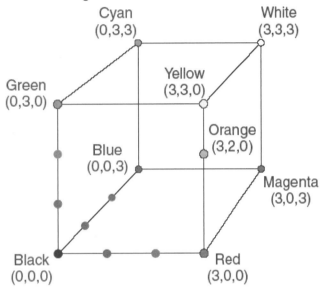

Figure 3.8 Three dimensional color cube

Textbook Reading

*   This lab appears on page **141** of the *Engineering Our Digital Future* textbook.
*   Prerequisite textbook reading before completing this lab: pp. **137-140.**

Engineering Designs and Resources

Worksheets used in this lab:

*   **L03-04-01 Color Representation Manual.Lst:** Students control the value of each of the three color components and view the resulting color value assigned.
*   **L03-04-02 Color Representation Automatic.Lst:** Students observe average color of a camera image.

3.4.1   Color Representation - Manual

Worksheet Description

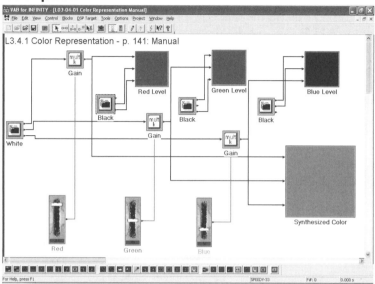

Figure 3.9 Color Representation Color Matcher

Open the worksheet *L03-04-01 Color Representation Manual.Lst*. You should see a screen that looks like Figure 3.9.

Three sliders allow you to individually control the amount of red, green and blue light that will be combined to make the synthesized color in the lower right portion of the worksheet. The three smaller displays at the top of the worksheet show the amount of each color component that you have selected with the sliders. Each slider can create values between 0 and 255 to represent color components specified by 8 bits.

Two small 100x100 pixel input images are used to create the color components. One is a completely white image. With 8 bits per pixel for each color component of the white image, all pixels have red = blue = green = 255. The other is a black image in which all pixels have a value of red = green = blue = 0.

The four image displays each show images that are all one color. The color combinations are made by scaling the white image that is read in on the left side of the worksheet. The red component of the white image (which is exactly the same as the blue and the green components) is the input to an image multiplier block. The function of the multiplier block is to multiply the input image by a value. The output of the multiplier block is connected to the red input of the red level display and to the red input of the synthesized color display. Since the red level display shows only the red component, its green and blue display inputs are connected to a black input image.

The red slider should be set to 0 to have no red component in the synthesized image, and it should be set to 255 to have the maximum red value. The output of the slider is scaled by 1/255 so that its maximum value for multiplying the white image is 1.

Exactly the same set of blocks and computation are used to create the blue component and the green component of the synthesized color. We really only need one black image to provide image components that are all 0 for all three of the individual color displays, but three block images are used here to emphasize the independence of the color components.

Although we talk about adding the three components, there is no adder on this worksheet. Each component connected to the synthesized color display creates light of that color, and the light from all three is added by the eyes of the viewer.

Laboratory Procedures

Steps	Instructions
Step 1:	Right click the red slider and verify that the range of the slider is 0 to 255. Verify that the slider output is multiplied by 1/255 using an "Ax + b" conversion.
Step 2:	Start your worksheet. Set all the slider values to 0.
	Q1: What do you see in the four displays?
Step 3:	With the blue and green sliders set to 0, slide the red slider through its full range of values.
	Q2: What do you see in the four displays? Describe the path you are following in a color cube when you move the red slider with the other two components set to 0. Draw a color cube like the one shown in the Figure 3.7 above, but use a maximum value of 255 instead of 3. Draw the line that represents the path you followed.
Step 4:	With the blue and red sliders set to 0, slide the green slider through its full range of values.
	Q3: What do you see in the four displays? Describe the path you are following in a color cube when you move the green slider with the other two components set to 0.
Step 5:	With the green and red sliders set to 0, slide the blue slider through its full range of values.

Steps	Instructions (Continued)
	Q4: What do you see in the four displays? Describe the path you are following in a color cube when you move the blue slider with the other two components set to 0.
Step 6:	Set all the sliders to the maximum value.
	Q5: What do you see in all four displays?
Step 7:	With the blue and green sliders set to the maximum value, slide the red slider through its full range of values.
	Q6: What do you see in the four displays? Describe the path you are following in a color cube when you move the red slider with the other two components set to 255.
Step 8:	With the blue and red sliders set to the maximum value, slide the green slider through its full range of values.
	Q7: What do you see in the four displays? Describe the path you are following in a color cube when you move the green slider with the other two components set to 255.
Step 9:	With the red and green sliders set to the maximum value, slide the blue slider through its full range of values.

Steps	Instructions (Continued)
	Q8: What do you see in the four displays? Describe the path you are following in a color cube when you move the blue slider with the other two components set to 255.
Step 10:	Adjust the sliders to make the following colors and record your results. For each color, predict the RGB components on a color cube where you would expect to find it. After you have synthesized the color, record the actual red, green and blue values. The RGB color representation is not very intuitive for most people and it is often surprising to see where a color is actually located.

	Your Prediction			Actual		
Color	Red Value	Green Value	Blue Value	Red Value	Green Value	Blue Value
Gray						
Orange						
Very light green						
Very dark red						
Brown						
Your favorite color						

Step 11:	Stop the worksheet. Close the worksheet.

3.4.2 Color Representation - Automatic

Worksheet Description

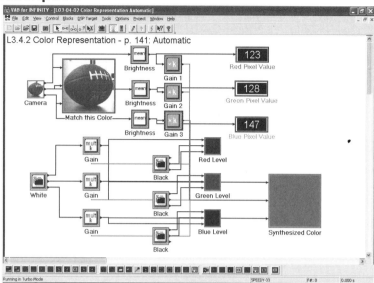

Figure 3.10 Color Representation - Automatic

Open the worksheet *L03-04-02 Color Representation Automatic.Lst.* You should see a screen that looks like Figure 3.10.

The worksheet creates a synthesized color that will match the average value of the colors seen by a camera. If an object is placed close enough to the camera to fill its field of view, you can use this worksheet to find the color components for that particular object color. Knowing the red, green, and blue component values would allow you to recreate that particular color and brightness level whenever you wanted to use it.

The only difference between this and the previous laboratory is that here the values of the three color components are controlled by the outputs of the three mean value blocks instead of three sliders. Each color component is connected to the input of a mean value block. The output of the mean value block is shown in a numeric display. In the example shown in Figure 3.10 the mean value blocks show 147 for blue, 128 for green, and 123 for red. The image contains a blue toy football on a wooden surface with a light background. The red and green values are higher than you might expect for a scene with no red or green objects because white has large red, green, and blue components.

Laboratory Procedures

Steps	Instructions
Step 1:	Start the worksheet.
Step 2:	Put an object with a well defined color close to the camera so that the camera only sees the object. You could choose a sweater or a notebook for example.

Steps	Instructions (Continued)
	Q1: How well does the synthesized color match the main color you perceive in the input image?
	Q2: Write the color name and also record the average red, green, and blue values.
Step 3:	Move the same object away from the camera so that it only takes up about half of the field of view like the football in the Figure 3.10.
	Q3: How did the average values of the three components change? Record the new values and explain why they are different from the previous values.
Step 4:	Place an object with many bright colors in front of the camera. You might try part of the comics from the Sunday newspaper for example. Record the average red, green and blue values.
	Q4: Are these values representative of any particular color in the object? How do you explain these results.
Step 5:	Repeat Steps 2 and 3 for several other objects.
Step 6:	Stop the worksheet. Close the worksheet.

Optional Further Explorations: What is actually being averaged?

Viewing the red, green, and blue components separately would make it easier to understand how the average values are related to the content of the color image. Remember that we have been using the green color component of a color image to be a grayscale image in previous laboratories, so we can also do the same with the red and blue component.

Steps	Instructions
Step 1:	Open the worksheet *L03-04-02 Color Representation Automatic.Lst.*
Step 2:	Click the "match this display" image display to select it. Type **Control C** to copy it and type **Control V** three times to paste three new displays on your worksheet.
Step 3:	Drag each of the new displays over to the right side of the worksheet so they all will be seen at the same time. You can drag the numeric displays to the left to make more room.
Step 4:	Right click the labels of the new displays and rename them "red component", "green component", and "blue component".
Step 5:	Connect the new displays: Click the **connection tool** icon at the top of the worksheet. (It looks like a horizontal line connecting two points.) • Connect the top (red) camera output to the three inputs of the "red component" image display. • Connect the middle (green) camera output to the three inputs of the "green component" image display. • Connect the lower (blue) camera output to the three inputs of the "blue component" image display. After making all the connections, click on the **set-up tool** icon at the top of the worksheet. (It is the diagonal arrow next to the connection tool.) Your worksheet should look like Figure 3.11. 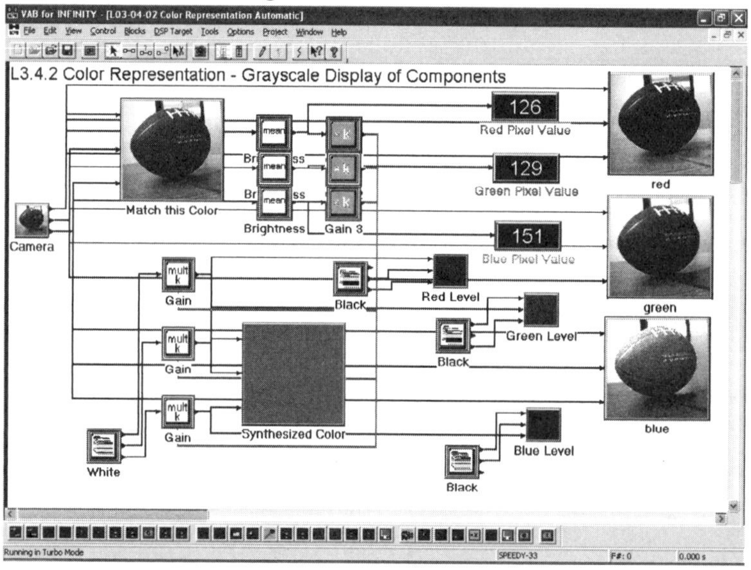

Figure 3.11 Color Representation - Grayscale Display of Components.

Steps	Instructions (Continued)
Step 6:	Start the worksheet. You will see three additional grayscale images that correspond to the values of the three color components. The numeric display labeled Red Pixel Value is the average intensity of the display you have labeled "red component".
Step 7:	Put an object with a well defined color close to the camera so that the camera only sees the object.
	Q5: Do the three color component displays have similar images? Explain the difference you see. Do these differences explain the three average values in the numeric displays?
Step 8:	Move the same object away from the camera so that it only takes up about half of the field of view like the football in the figure above.
	Q6: Do the three color component displays have similar images? Explain the difference you see. Do these differences explain the three average values in the numeric displays?
Step 9:	Find (or make) part of a page or object that has black text on a bright red background. Place that object in front of the camera.
	Q7: Can you read the black text easily in the red component image? Explain why it is more difficult to read the text in the blue or green component images.
Step 10:	Experiment with black text and white text on brightly colored backgrounds of different colors and note which color component images are easiest to read.
Step 11:	Stop the worksheet. You will use this worksheet in the Optional Further Explorations section below, so you may wish to save it as *Mylab4* on your desktop using the "Save As" option under File on the top toolbar.

Optional Further Explorations: Color display of components

Viewing the red, green, and blue components as gray scale images in the previous exploration helped us understand how the average values were related to the image content. It would be more interesting to display them in the colors that are added to make the color display.

Steps	Instructions
Step 1:	Start with the worksheet used in the previous exploration, and make some different connections
	Click the **deletion tool** icon at the top of the worksheet. (It looks like a pair of scissors.)
	• Disconnect the green and blue inputs to the red component display by clicking them with the left mouse button.
	• Disconnect the red and blue inputs to the green component display.
	• Disconnect the red and green inputs to the blue component display.
Step 2:	Click the **connection tool** icon at the top of the worksheet. (It looks like a horizontal line connecting two points.)
	• Connect an output from any of the three black images (it does not matter which one) to the following six inputs:
	-the green and blue inputs of the red component display
	-the red and blue inputs of the green component display
	-the red and green inputs to the blue component display
	After making all the connections, click on the **set-up tool** icon at the top of the worksheet. (It is the diagonal arrow next to the connection tool.)
	Your worksheet should look like Figure 3.12.
	 Figure 3.12 Color Representation - Color Display of Components
Step 3:	Start the worksheet.
	The three additional displays that showed the grayscale images of the color components will now show the correct colors.

Steps	Instructions (Continued)
Step 4:	Find part of a page or object that has black text on a bright red background. Place that object in front of the camera.
	Q8: How do red parts of the image appear in each of the three component displays? How do black parts of the image appear in each of the three component displays?
Step 5:	Find an object or page with many brightly colored areas and view it with the camera.
	Q9: Make a table that lists the colors of the areas and indicate whether the areas are bright, medium, or dark in the three color component displays.

Color	Red Display	Green Display	Blue Display
1.			
2.			
3.			
4.			
5.			
6.			

Steps	Instructions (Continued)
	Q10: Compare your results in Question 9 to your plots of colors on the color cube in Lab 3.4.1
Step 6:	Stop the worksheet. Close the worksheet.

Summary

We can make almost all the colors we normally see by adding the right proportions of red, green, and blue light.

3.5 Resolution Trade-offs

Lab Objectives

In Lab 3.1 and 3.2 we explored the effects of changing quantization, which determines the number of bits per pixel, and changing the sampling interval, which changes the number of pixels in an image. Reducing either the number of pixels or the number of bits per pixel reduces the storage required for an image at a cost of lowered image quality. In this laboratory we will explore how to stay within a "bit budget" for an image by adjusting both the quantization levels and the sampling to keep the best image quality. This can be important when deciding what images will be put on a web page or sent electronically.

Textbook Reading

- This lab appears on page **151** of the *Engineering Our Digital Future* textbook.
- Prerequisite textbook reading before completing this lab: pp. **143-150.**

Engineering Designs and Resources

Worksheets used in this lab:

- **L03-05-01 Resolution Trade-Offs Gray Scale.Lst**: Students evaluate the visual impact of changing the number of bits per pixel and/or bit resolution on a grayscale image.
- **L03-05-02 Resolution Trade-Offs Color.Lst**: Students evaluate the visual impact of changing the number of bits per pixel and/or bit resolution on a color image.

3.5.1 Resolution Trade-Offs - Gray Scale

Worksheet Description

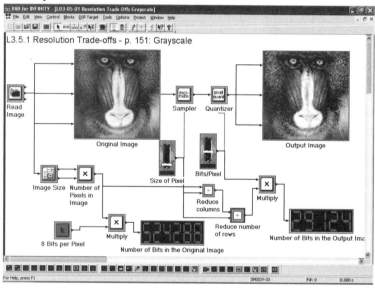

Figure 3.13 Resolution Trade-offs Gray Scale

Open the worksheet *L03-05-01 Resolution Trade-Offs Gray Scale.Lst.* You should see a screen that looks like Figure 3.13 above except that the image and numeric displays will be blank.

An input image is read by the read image block on the left. The output of the read image block is connected to the three inputs of the "Original Image" display and also to the input of a decimate block. As we learned in Lab 3.2, this block will produce an output image with a reduced number of pixels. The output of the decimate block is connected to the input of a pixel quantizer block. As we learned in Lab 3.1, the quantizer reduces the number of bits used for each pixel. The image output of the pixel quantizer block is connected to the inputs of the "Output Image Display". By observing this display we can see the effect on the appearance of the image when we reduce both the number of pixels and the number of bits per pixel. Two sliders control the parameters for the decimate block and the quantizer block. The slider labeled "size of pixel" in this worksheet is controlling the decimate block in the same way as the slider labeled "Downsampling Factor" did in Lab 3.2.

Laboratory Procedures

Steps	Instructions
Step 1:	Start the worksheet.
Step 2:	Set the pixel size slider to 1 and the bits per pixel slider to 8. Verify that the image of the baboon on the right is the same as the image on the left and that the same number of bits is used for each image.
Step 3:	Set the pixel size to 2.
	Q1: How many bits are now used in the output image? What is the ratio of bits in the output image to bits in the input image?
	Q2: What parts of the output image look the same as the original image? What parts look different? How would you characterize the parts of the image that show the most loss of quality?
Step 4:	Return the pixel size to 1. Adjust the bits per pixel until you have an image using the same number of bits that you recorded in Q1.
	Q3: How many bits per pixel are used?

Steps	Instructions (Continued)
	Q4: What loss of image quality do you observe? Which image is better - this one or the one from Step 3?
Step 5:	Set the pixel size to 2 and set the bits per pixel to 8. Slowly reduce the number of bits per pixel until you just notice a further reduction in image quality.
	Q5: How many bits per pixel are you using? How many bits are needed for the image?
	Q6: For the following bit budgets, find the choices for each slider that will give you the best image quality. In most cases the best quality will not be very good.

Maximum bits in image	pixel size	bits per pixel
300,000		
200,000		
150,000		
100,000		
75,000		
50,000		

Steps	Instructions (Continued)
	Q7: In this worksheet we are downsampling first and then we are reducing the number of bits per pixel in the downsampled image. Would our result be any different if we reduced the number of bits first and then downsampled?
Step 6:	Now we will look at a different image. With the right mouse button, click the **Read Original Image Block**. Click the **Browse** button and then navigate to Program Files\Hyperception\VAB-INF\Images\BWimages and select *cityLights256x256.bmp* in this directory. Click **Open** in the browser's Open window, and then click **OK** in the Bitmap Read Parameters Window
Step 7:	Compare the output image with a pixel size of 2 and 8 bits per pixel to the output image with a pixel size of 1 and 2 bits per pixel.
	Q8: Which image has the better image quality? Is your choice different from the baboon image? Why?
	Q9: Repeat Question 6 for this image.

Maximum bits in image   **pixel size**   **bits per pixel**

300,000

200,000

150,000

100,000

75,000

50,000

Steps	Instructions (Continued)
	Q10: Why would you spend your bit budget differently on the *cityLights* image than on the baboon image?
Step 8:	Change the input image to the *girlOnWater256x256.bmp* image.
Step 9:	Explore the effects of changing the pixel size and the number of bit per pixel.
	Q11: Find the best settings for an image with less than 25,000 bits. What value of pixel size and bits per pixel did you use? Would this image be suitable for a small part of a web page? Would this image be suitable for a full screen display?
Step 10:	Change the input image to the *canoeBW512x512.bmp* image. Use the same settings for the sliders that were used in Question 11.
	Q12: How does this output image compare to the previous output image in quality? How many bits are in this output image?
	Q13: What happens if you try to reduce this image to less than 25,000 bits?
	Q14: Why is the canoe image with less than 25,000 bits so much poorer in quality that the *girlOnWater* image? What characteristics of the image content of the canoe image require more bits to be reasonably reproduced in the output image?

Steps	Instructions (Continued)
Step 11:	Stop the worksheet. Close the worksheet.

3.5.2 Resolution Trade-Offs: Color

Worksheet Description

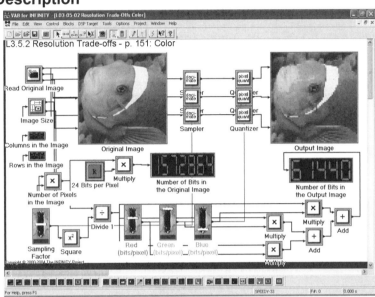

Figure 3.14 Resolution Trade-Offs: Color

Open the worksheet *L03-05-02 Resolution Trade-offs Color.Lst.* You should see a screen that looks like Figure 3.14 above except that the image and numeric displays will be blank.

This worksheet is very similar to the previous worksheet except for changes needed to process and display color images instead of grayscale images. The image read block has three outputs for the red, green, and blue color components. Each component is individually decimated and quantized in the same manner as the previous laboratory. The same decimation factor is used by all three color components, but the number of bits per pixel can be individually set for each color.

Laboratory Procedures

Steps	Instructions
Step 1:	Start the worksheet.
Step 2:	Set the sampling factor to 1 and all three sliders for bits per pixel to 8.
Step 3:	Adjust the sliders to get the best quality image you can get with less than 250,000 bits.

Steps	Instructions (Continued)
	Q1: What values give the best quality? **Sampling** **bits per pixel** **Factor** **red** **green** **blue**
Step 4:	Adjust the sliders to get the best quality image you can with less than 200,000 bits.
	Q2: What values give the best quality? **Sampling** **bits per pixel** **Factor** **red** **green** **blue**
Step 5:	Increase the sampling factor to 4, and set all three color components to 8 bits per pixel.
	Q3: What happens to the image quality? Specifically look at the eye of the clown-fish and the boundaries of the white band.
	Q4: How many bits are required for this image?
Step 6:	Find the best image using less than 100,000 bits with a sampling factor of 2. Compare the quality of the output image with the output image from Step 5.
	Q5: What settings did you use for the number of bits per pixel for each color component?

Steps	Instructions (Continued)
	Q6: Which image do you prefer and why?
	Q7: How does your evaluation of the relative quality of the two images change if you view them from a distance of ten feet instead of from very close to the display screen?
Step 7:	Use the *canoe512x512.bmp* which can be found in the Program Files/Hyperception\VABINF\Images\colorImages directory.
	Q8: What are the best settings for this image if it must be stored using less than 250,000 bits?
	Q9: What is the ratio of the number of bits in Question 8 to the number of bits in the original image?
Step 8:	Explore these effects on other color images and determine what kinds of images can have the number of bits in the image reduced with the least impact on image quality.
Step 9:	Stop the worksheet. Close the worksheet.

Overview Questions

A: Why do we use different numbers of bits for the red, green, and blue color components but use the same "size of pixel" for all three color components? What would happen to colors in the image if we had different pixel sizes for each color component?

B: How can we reduce the number of bits in an image without reducing the number of bits/pixel or increasing the pixel size when there is a lot of background that is not interesting to us? (Hint: Think of the three types of lenses discussed in Chapter 3.)

Summary

The number of bits used for an image can be reduced in two ways. The number of bits per pixel can be reduced and the total number of pixels in the image can be reduced. The amount of reduction that can be made without unacceptable loss of quality depends on the content of the image.

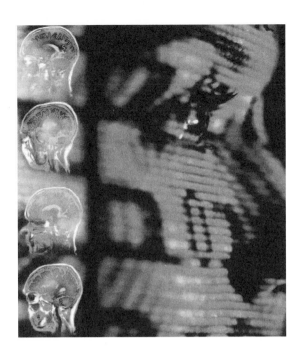

Math You Can See

In this chapter we will explore how we can use digital images. We will learn a number of digital darkroom techniques that can improve a digital image and we will learn how to find information in an image by applying mathematical operations. In Chapter 3 we focused on creating digital images that had the right number of colors and the right number of pixels. Now we will consider an image as a matrix of numbers so we can use simple mathematical operations to change images and combine images.

Infinity Labs

Introduction

The laboratories in Chapter 4 will focus on changing images to improve their appearance or make them more useful for applications. When an image or video sequence is stored in digital form, there is always a trade-off between the storage cost in terms of number of bits used and the quality of the stored images. Chapter 3 explored several aspects of digital image representation. Those laboratories demonstrated the effects of reducing the number of bits used for storage by decreasing the number of bits in each pixel, decreasing the number of pixels in each image, and decreasing the number of images in a video sequence. In this chapter we will assume that the quality of the images is good so that computational methods can be used to interpret the content of the images. For example, we may want to find edges of objects in an image or find all red objects in an image or set off an alarm when something in an image sequence moves.

The basic image input and output VAB blocks used in Chapter 3 will also be used in this chapter. New blocks will be added to change pixel values in an image to correct brightness and contrast. Blocks to add and subtract images can compare images taken at different times or different parts of the same image. These operations will be ordinary simple operations from arithmetic that are used on the arrays of numbers that represent an image. The main difference from simple arithmetic is that image values will have a limited range, usually 0 to 255, and VAB blocks will offer choices for what to do with results of arithmetic that fall outside that range.

4.1 Brightness and Contrast

Lab Objectives

Two of the most basic digital darkroom operations for improving images are adjusting the brightness and adjusting the contrast to make an image more visually pleasing. In this laboratory we will explore the simple arithmetic operations that change brightness and contrast, and we will see how these operation are affected by the limited number of values that a pixel can have. We will also look at what happen when both brightness and contrast need adjustment. The different effects of adjusting brightness first or adjusting contrast first demonstrate the distributive law in algebra.

Textbook Reading

- This lab appears on page **205** of the *Engineering Our Digital Future* textbook.
- Prerequisite textbook reading before completing this lab: pp. **187-204.**

Engineering Designs and Resources

Worksheets used in this lab:

- **L04-01-01 Brightness and Contrast Grayscale.Lst**: Students change brightness and contrast of a grayscale image using multiply and add operations.
- **L04-01-02 Brightness and Contrast Color Image.Lst:** Students change brightness and contrast of a color image using multiply and add operations.
- **L04-01-03 Brightness and Contrast Color Camera.Lst**: Students change brightness and contrast of a color image from a camera using multiply and add operations.
- **L04-01-04 Brightness and Contrast Cascade.Lst**: (Students change both brightness and contrast of a color image and observe the effect of the order of the operations.

4.1.1 Brightness and Contrast - Grayscale

Worksheet Description

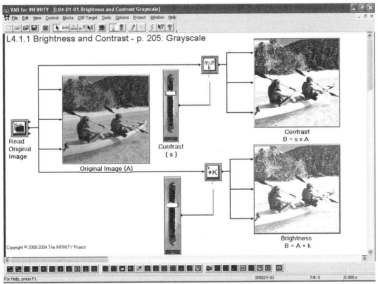

Figure 4.1 Brightness and Contrast - Grayscale

Open the worksheet *L04-01-01 Brightness and Contrast Grayscale.Lst.* You should see a screen that looks like Figure 4.1 above except that the image and numeric displays will be blank.

The worksheet has three image displays so that the images after contrast adjustment or brightness adjustment may be compared to each other and to the original image. On the left side of the worksheet a block reads an image file. The middle image output from this block is used as the grayscale output. If the image read is actually a grayscale image, then it does not matter which output we use because all three will all be the same. If the image read is a color image, then the green output is usually a good approximation to a grayscale image. This middle output from the image file reading block is connected to all three inputs of the original image display. The same image output is also connected to the inputs of two arithmetic processing blocks which change contrast using multiplication or change brightness using addition. Both blocks have a parameter controlled by a slider. The output of each of the blocks is connected to the inputs of a display block.

Running Your Worksheet

Steps	Instructions
Step 1:	Start the worksheet.
Step 2:	Set the contrast slider to 1 and the brightness slider to 0.
	Q1: Do you expect both images to be the same? Explain why.
Step 3:	Slowly change the contrast from 1 to 2 and observe the changes in the Contrast Image.
	Q2: What happens to this image as the contrast is increased?
	Q3: If s=2, what input pixel values will be converted to contrast image values that are greater than 255? How are values greater than 255 displayed in the Contrast image?
Step 4:	Find four areas of the canoe and the shore that are bright white in the Contrast image, and verify that they have a value of 255 by placing the cursor over them and clicking to read the value. (Note that the display is a general color display, so all color components will be the same for a grayscale image.

Steps	Instructions (Continued)
	Q4: For the four areas that have output pixels = 255, what are the values of the corresponding input pixels? Record these values. Is this consistent with your answer to Question 3?
Step 5:	Find two areas in the sky that are not bright white in the Contrast image, and record those pixel values.
	Q5: What are the values of the corresponding input pixels? Record these values. Is this consistent with your answer to Question 3?
Step 6:	Slowly change the brightness from 0 to 255 and observe the changes in the Brightness Image.
	Q6: How do you explain the Brightness image when k=255
Step 7:	Change the brightness to a value close to 80.
	Q7: If k=80, what input pixel values will be converted to brightness image values that are greater than 255? How are values greater than 255 displayed in the Brightness image?
Step 8:	Find four areas of the canoe and the shore that are bright white in the Brightness image, and verify that they have a value of 255 by placing the cursor over them and right clicking to read the value.

Steps	Instructions (Continued)
	Q8: For these four areas, what are the values of the corresponding input pixels? Record these values. Is this consistent with your answer to Question 7?
Step 9:	Adjust the contrast image so that the overall light level of the contrast image best matches the overall light level of the brightness image with k approximately 80. This will be a judgment call because there will not be a perfect match.
	Q9: What contrast value did you select? Compare this to the values selected by the other students in the class. There is no "right" answer because the images will never look the same, so it is expected that there will be many different answers.
	Q10: Compare the appearance of the two images.
	Q11: For the following pixel values indicate what value the contrast image and brightness image would have at your selected settings of s and k in Step 9.

<table>
<tr><td></td><td style="text-align:center">s =</td><td style="text-align:center">k =</td></tr>
<tr><td>input pixel</td><td>contrast image value</td><td>brightness image value</td></tr>
<tr><td style="text-align:center">0</td><td></td><td></td></tr>
<tr><td style="text-align:center">10</td><td></td><td></td></tr>
<tr><td style="text-align:center">50</td><td></td><td></td></tr>
<tr><td style="text-align:center">100</td><td></td><td></td></tr>
<tr><td style="text-align:center">150</td><td></td><td></td></tr>
<tr><td style="text-align:center">200</td><td></td><td></td></tr>
<tr><td style="text-align:center">255</td><td></td><td></td></tr>
</table>

Steps	Instructions (Continued)
Step 10:	Slowly change the contrast from 1 to 0.5 and observe the changes in the Contrast Image.
	Q12: Describe the changes in the image.
	Q13: With a contrast setting of s=0.5, what is the largest value a contrast image pixel can have?
Step 11:	Slowly change the contrast from 0.5 to 0.0.
	Q14: Describe the changes in the image. What happens when s=0? Why?
Step 12:	Slowly change the brightness from 0 to -255 and observe the changes in the Brightness Image.
	Q15: How do you explain the brightness image when k = -255? How is this different from what happens when s=0?
Step 13:	Set the brightness to a value close to -80. Adjust the contrast so that the overall lightness of the two images looks the same.
	Q16: What value did you select for contrast? Compare this to the values selected by the other students in the class.

Steps	Instructions (Continued)
	Q17: Compare the appearance of the two images.
	Q18: For the following pixel values indicate what value the contrast image and-brightness image would have at your selected settings of s and k in Step 13. s = k = **input pixel** **contrast image value** **brightness image value** 0 10 50 100 150 200 255
Step 14:	Stop the worksheet. Close the worksheet.

4.1.2 Brightness and Contrast - Color Image

Worksheet Description

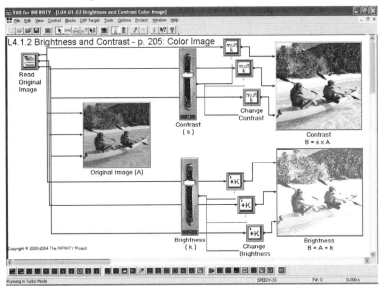

Figure 4.2 Brightness and Contrast - Color Image

Open the worksheet *L04-01-02 Brightness* and *Contrast Color Image.Lst.* You should see a screen that looks like Figure 4.2 above except that the image and numeric displays will be blank.

Like the previous worksheet it has three image displays to show the original image, the contrast adjusted image, and the brightness adjusted image. There are three multiplication blocks to adjust the contrast for the three color components. The same slider controls the scaling parameters for all three blocks. There are also three addition blocks to adjust the brightness for the three color components. A single slider controls the scaling parameters for all three blocks.

Laboratory Procedures

Steps	Instructions
Step 1:	Start the worksheet.
Step 2:	Set the contrast to 1.0 and the brightness to 0.0.
	Q1: Are the Contrast image and Brightness image the same as the Input image? Why?
Step 3:	Slowly change the contrast from 1 to 2 and observe the changes in the Contrast Image.

Steps	Instructions (Continued)
	Q2: Describe the changes in the Contrast image.
	Q3: Is it possible that changing the contrast can change the color you perceive if the perceived color is related to the relative proportion of red green and blue? What could cause the perceived color to change?
	Q4: Read the pixel values for the dark blue part of the sky, and then read the pixel values for the same part of the sky in the contrast image. Record these values. Why does the color look different?
Step 4:	Adjust the brightness image so that the overall light level of the brightness image best matches the overall light level of the contrast image with s = 2. This will be a judgment call because there will not be a perfect match.
	Q5: What value did you select for the brightness? Compare this to the values selected by the other students in the class.
	Q6: Compare the appearance of the two images.

Steps	Instructions (Continued)
	Q7: For the following color component values (red, green, or blue) indicate what value the contrast image and brightness image would have for the color component at your selected settings of s and k in Step 4. input color component value contrast image value (s) brightness image value (k) 0 10 50 100 150 200 255
	Q8: For the following pixel values indicate what value the contrast image and brightness image would have. Determine what the color of each image would be. (s) (k) input pixel contrast image value brightness image value R G B R G B R G B 200 100 0 100 50 0
Step 5:	Set the contrast to a value close to 0.5. Adjust the brightness image so that the overall light level of the brightness image best matches the overall light level of the contrast image with s approximately equal to 0.5. This will be a judgment call because there will not be a perfect match.
	Q9: What value did you select for the brightness? Compare this to the values selected by the other students in the class.

Steps	Instructions (Continued)
	Q10: Compare the appearance of the two images.
Step 6:	Stop the worksheet. Close the worksheet.

4.1.3 Brightness and Contrast - Camera

You may continue experiments with brightness and contrast adjustments on other images by selecting new file names in the "read original image" block.

You may also find it interesting to adjust the brightness and contrast of images from the camera by opening worksheet *L04-01-03 Brightness and Contrast Color Camera.Lst*. This worksheet is exactly like the previous worksheet in Lab 4.1.2 except that the image file reading block is replaced with a camera block.

4.1.4 Brightness and Contrast - Cascade

Worksheet Description

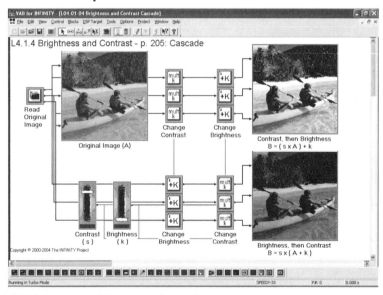

Figure 4.3 Brightness and Contrast - Cascade

Open the worksheet *L04-01-04 Brightness and Contrast Cascade.Lst*. You should see a screen that looks like Figure 4.3 above except that the image and numeric displays will be blank. It is similar to the worksheet for Lab 4.1.2 except that each color component has both contrast and brightness adjustments. In the upper display the contrast is adjusted first by multiplication and then the brightness is adjusted by addition. In the lower display the brightness is adjusted first and then the contrast adjustment is made. One slider controls the scale factor s for all contrast adjustment of all color components. A second slider controls the constant k added for brightness adjustments for all color components.

Laboratory Procedures

Steps	Instructions
Step 1:	Start your worksheet.
Step 2:	Adjust the contrast and brightness sliders and note that the two adjusted images do not look the same (Although they do not look dramatically different.)
	Q1: If s = 1, are the two adjusted images the same when the brightness is adjusted? Why?
	Q2: If k = 0, are the two adjusted images the same when the contrast is adjusted? Why?
Step 3:	Set s = 1.5 and k = 50. If your slider will not give these values exactly, pick values close to them.
	Q3: Which image seems lighter? Why? (Hint: When the multiplication is done second, the scale factor s multiplies both the pixel value and the added constant.)
Step 4:	Set s = 1.5 and k = -50. If your slider will not give these values exactly, pick values close to them.
	Q4: Which image seems lighter? Why?

Steps	Instructions (Continued)
Step 5:	Set the contrast using s = 1.8. Adjust the brightness slider until you think the quality of the upper image is best. Record this value. Then adjust the brightness slider until you think the quality of the lower image is best. Record this value.
	Q5: How are the two brightness values related?
	Q6: Determine what values s and k should have if we want the upper image display to show the input pixel range from 64 to 191 spread out over the whole output pixel range of 0 to 255.
	Q7: Repeat Question 6 for the lower display.
	Q8: How are the constants from Steps 5 and 6 related?
Step 6:	Stop the worksheet. Close the worksheet.

Overview Questions

A: When an image is brightened or darkened, why is the possible range of pixel values smaller than the original range? If brightening improves the appearance, what can we say about the actual range of the pixel values in the original image?

B: Usually both contrast and brightness are adjusted on the same image. If contrast is increased and then often brightness is reduced. If this improved an image by spreading the pixel values over the full range from 0 to 255, what would the range of the original pixel values be?

C: If both contrast and brightness adjustments are made of the same image, does it matter what order of operations is used? For the same final result compare the values of brightness that would be used if brightness were adjusted in the second operation instead of the first. How is this affected by the limited range of 0 to 255?

Summary

A digital image is an array of numbers. Digital darkroom effects are created by mathematical operations on these numbers. If we add a constant value to all the numbers in an image, it will appear brighter or darker when we view it. If we multiply all the numbers in an image by a constant value, it will appear to have more or less contrast when we view it. Usually both brightness and contrast are adjusted to improve the appearance of an image.

4.2 Threshold and Negation

Lab Objectives

In this laboratory we will explore the use of blocks which change pixel values. In the previous laboratory we changed pixel values by adding a constant to brighten or darken and image. We also multiplied pixels by a constant to change the contrast. Now we will use blocks that create a negative of the image or threshold an image. Then we will look at more general ways to change pixel values that are defined by functions that we can draw graphically. We can associate an output pixel value with an input pixel value either by making a table of all the (input value, output value) pairs or by drawing the graph of the table. We will see that drawing the graph of the output pixel value as a function of the input pixel value is much easier than making a long table.

Textbook Reading

* This lab appears on page **209** of the *Engineering Our Digital Future* textbook.
* Prerequisite textbook reading before completing this lab: pp. **205-209**.

Engineering Designs and Resources

Worksheets used in this lab:

* **L04-02-01 Threshold and Negation Grayscale Image.Lst:** Students observe negative image and adjust threshold for a grayscale image.
* **L04-02-02 Threshold and Negation Grayscale Camera.Lst:** Students observe negative image and adjust threshold for a grayscale image from a camera.
* **L04-02-03 Threshold and Negation Color.Lst:** Students observe negative image and adjust threshold for a color image.
* **L04-02-04 Threshold and Negation Pixel Map Grayscale.Lst:** Students use graphical input for general mapping for a grayscale image.
* **L04-02-05 Threshold and Negation Pixel Map Color.Lst:** Students use graphical input for general mapping for a color image
* **L04-02-06 Threshold and Negation Pixel Map Color Camera.Lst:** Students use graphical input for general mapping for a color image from a camera.

4.2.1 Threshold and Negation - Grayscale Image

Worksheet Description

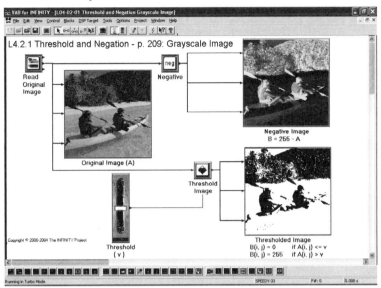

Figure 4.4 Threshold and Negation - Grayscale Image

Open the worksheet *L04-02-01 Threshold and Negation Grayscale Image.Lst.* You should see a screen that looks like Figure 4.4 above except that the image displays will be blank.

An input image is read into the worksheet using a Bitmap Read block and this image is displayed as the Original Image. The output of the Bitmap Read block is also connected to the input of a Negative block and the input of a Threshold Image block. The Negative Image display shows the output of the Negative block. The Threshold Image block uses a threshold level set by the slider. The output of this block is displayed in the Thresholded Image display.

Laboratory Procedures

Steps	Instructions
Step 1:	Right click the slider and set the number of steps to 256.
Step 2:	Start the worksheet. Observe the negative image. Note that the lightest areas of the original image are the darkest areas of the negative image.

Steps	Instructions (Continued)
	Q1: Record five pixel values on the brightest part of the canoe in the original image by placing the cursor over each of the five pixels and clicking. Record five values in the same part of the canoe on the negative image. Are these values consistent with the equation for the negative B = 255 - A? (Note that it is not possible to get exactly the same pixel positions in both images, but, since the side of the canoe does not have much intensity variation, the values should be close to what is predicted by the equation.)
Step 3:	Set the Threshold slider value to 128 and observe the thresholded image.
	Q2: Based only on the thresholded image, what can you say about the pixel values on the side of the canoe? On the life jacket? In the sky? At the ends of the paddle?
Step 4:	Increase the Threshold slider value to 170 and observe the thresholded image.
	Q3: Did the number of black pixels increase or decrease when the threshold value was increased? Why?
	Q4: Repeat question 2 at this threshold value.
Step 5:	Set the Threshold slider value to 20 and observe the thresholded image.

Steps	Instructions (Continued)
	Q5: What parts of the image have pixel values less than 20?
Step 6:	Vary the threshold value to answer the following questions.
	Q6: What is the darkest pixel in the sky? Where is it located?
	Q7: What is the brightest pixel on the side of the canoe? Where is it located?
Step 7:	Stop the worksheet. Close the worksheet.

You may continue experiments with threshold adjustments and negation on other images by selecting new file names in the "read original image" block.

4.2.2 Threshold and Negation Grayscale - Camera

Worksheet Description

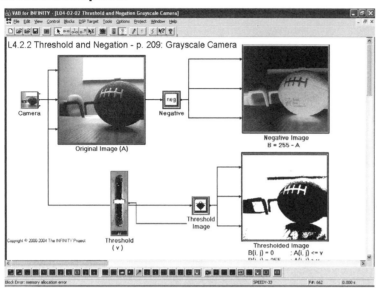

Figure 4.5 Threshold and Negation Grayscale - Camera

You may also find it interesting to explore threshold adjustments and negation on grayscale images using a single color component from the camera by opening worksheet *L04-02-02 Threshold and Negation Grayscale Camera.Lst*. This worksheet is exactly like the previous worksheet in Lab 4.2.1 except that the image file reading block is replaced with a camera block.

Laboratory Procedures

Steps	Instructions
Step 1:	Start the worksheet.
Step 2:	Aim the camera at a scene with several objects and observe what happens as the threshold is adjusted.
	Q1: Why does the thresholded image change a little even when the threshold is constant and objects in the scene are not moving? List several possible causes for this variation.
Step 3:	Stop the worksheet. Close the worksheet.

4.2.3   Threshold and Negation - Color

Worksheet Description

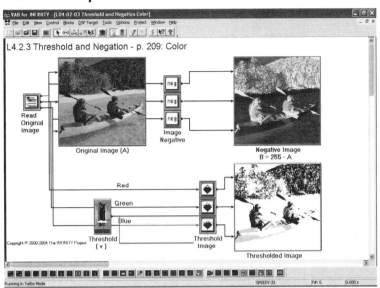

Figure 4.6 Threshold and Negation - Color

Open the worksheet *L04-02-03 Threshold and Negation Color.Lst.* You should see a screen that looks like Figure 4.6 above except that the image displays will be blank.

It is very similar to the worksheet from Laboratory 4.2.1. Since this worksheet uses color images, each of the three color components needs a Negative block and a Threshold Image block.

Laboratory Procedures

Steps	Instructions
Step 1:	Right click the slider and set the number of steps to 256.
Step 2:	Start the worksheet and observe the negative image.
	Q1: What are the "negative colors" for the following: -The bright yellow canoe -The blue life jacket -The red hat
Step 3:	Set the Threshold slider to a value of 138 and observe the thresholded image.
	Q2: Why can a thresholded image only have eight colors? List the eight possible colors.

Steps	Instructions (Continued)
	Q3: What colors do you see in the sky? How does this compare with the original image?
	Q4: What colors do you see in the canoe inside and outside? How does this compare with the original image?
	Q5: What colors is the water?
Step 4:	Increase the threshold from 138 to 186 and observe the change in the water.
	Q6: Describe how the colors of the water changed and explain why the changes happened.
Step 5:	Vary the threshold to answer the following questions.
	Q7: Estimate the highest value of any color component in the water.

Steps	Instructions (Continued)
	Q8: Estimate the lowest value of any color component in the water.
	Q9: Estimate the highest value of any color component on the sandy shore. Which color component has this highest value?
	Q10: Estimate the highest value of any color component in the cloud. Which color component has this highest value?
Step 6:	Stop the worksheet. Close the worksheet.

4.2.4   Threshold and Negation - Pixel Map Grayscale

Worksheet Description

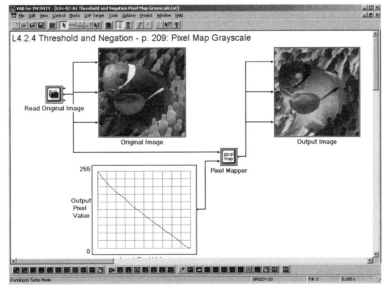

Figure 4.7 Threshold and Negation - Pixel Map Grayscale

Open the worksheet *L04-02-04 Threshold and Negation Pixel Map Grayscale.Lst.* You should see a screen that looks like Figure 4.7 above except that the image displays will be blank and the pixel map graph will show a horizontal line across the center.

Previous laboratories used a specific mapping function, such as a negative or a threshold, to create output pixel values. In this laboratory we will use a general Pixel Mapper block which uses the pixel map graph to associate an output pixel value with every input pixel value. The graph shows the output value as a function of the input pixel value.

Laboratory Procedures

Steps	Instructions
Step 1:	Start the worksheet.
Step 2:	Observe the pixel map graph and the Output Image.
	Q1: What does the pixel map graph tell you the output image values should be? Is this what the output looks like?
Step 3:	With the cursor, draw as straight a line as you can in the pixel map graph from the lower left corner to the upper right corner.
	Q2: Describe the output image. Explain how this pixel map converts input pixel values to output pixel values.
Step 4:	With the cursor, draw as straight a line as you can in the pixel map graph from the upper left corner to the lower right corner, as is shown in Figure 4.7 above.
	Q3: Describe the output image. Explain how this pixel map converts input pixel values to output pixel values.
Step 5:	With the cursor, draw a graph that is 0 when the input pixel value is less than 128 and is 255 when the input pixel value is greater than 128.

Steps	Instructions (Continued)
	Q4: Describe the output image. Explain how this pixel map converts input pixel values to output pixel values.
Step 6:	Draw a map that will produce an image that is black where the image from Step 5 is white and is white where the image from Step 5 is black.
	Q5: Sketch your graph here.
Step 7:	Draw a pixel map graph that brightens the image, and verify that the output image is brighter.
	Q6: Sketch your graph here.
Step 8:	Draw a pixel map graph that increases the contrast of the image, and verify that the output image has higher contrast.
	Q7: Sketch your graph here.
Step 9:	The graphs we have drawn above reproduce the effects and brightening, increasing contrast, thresholding, and creating a negative. However, any function can be used for a map. Draw a graph that looks like a capital W.
	Q8: Describe the output image for this map.

Steps	Instructions (Continued)
Step 10:	Draw a pixel map graph that you think creates the most interesting output image.
	Q9: Sketch your graph here. Describe the output image.
Step 11:	Stop the worksheet. Close the worksheet.

Optional Quantitative Investigations: What mapping functions have inverse functions?
The pixel map graph defines the output pixel value as a function of the input pixel value. For every input pixel value, exactly one output pixel value is specified. If the output of the pixel mapper is connected to the input of a second pixel mapper, we can visually show the result of a function of a function in a new display. In particular, if we can draw a second function that makes the new display look exactly like the original display, we will have drawn an inverse function. We will not be able to draw the pixel map graphs precisely enough to create perfect inverse functions, but we can get a good idea of what an inverse function would look like by creating a new display that is almost exactly like the original display.

Steps	Instructions
Step 1:	Open the worksheet *L04-02-04 Threshold and Negation Pixel Map Grayscale.Lst*.
Step 2:	Copy the pixel map graph by selecting it with a left mouse click. Then type **Control C.** Paste the copy onto the worksheet by typing **Control V.** In the same way, make a copy of the Pixel Mapper block and the Output Image block.
Step 3:	Arrange the worksheet so that the blocks do not overlap and the new Pixel Mapper block is just to the right of the first Pixel Mapper block. Right click the **Output Image** label of the new display and change it to "Output Image 2"

Steps	Instructions (Continued)
Step 4:	Connect the new blocks. Select the **Connect tool** icon from the toolbar. • Connect the output of the first Pixel Mapper block to the upper input of the second Pixel Mapper block. • Connect the output of the new pixel map graph to the lower input of the second Pixel Mapper block. • Connect the output of the second Pixel Mapper block to all three inputs of Output Image 2. Select the **Setup Tool** icon from the toolbar. Your worksheet should look similar to the worksheet shown below. 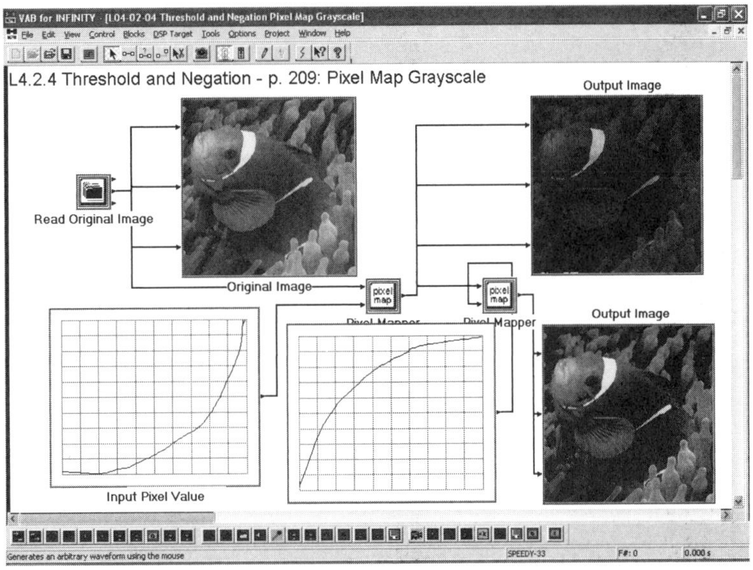 **Figure 4.8** Screen capture for Step 4
Step 5:	Start your worksheet.
Step 6:	With the cursor, draw as straight a line as you can in the first pixel map graph from the lower left corner to the upper right corner. Draw the same thing in the second pixel map graph.
	Q1: What do you see in the Output Image display and the Output Image 2 display?
Step 7:	With the cursor, draw as straight a line as you can in the first pixel map graph from the upper left corner to the lower right corner. Make no change in the second pixel map graph.

Steps	Instructions (Continued)
	Q2: Describe the Output Image display and the Output Image 2 display.
Step 8:	Draw a new map in the second pixel map graph that makes Output Image 2 look as much like the original image as possible.
	Q3: Sketch your second pixel map graph here. Describe the inverse function.
Step 9:	In the first pixel map graph, draw a graph that looks like $(x^2)/255$. Remember that the input values on the horizontal axis go from 0 to 255.
Step 10:	Draw a new map in the second pixel map graph that makes Output Image 2 look as much like the original image as possible.
	Q4: Sketch your second pixel map graph here. Describe the inverse function. Does it look like what you would have predicted using mathematics?
	Q5: Describe the Output Image.
Step 11:	Reverse the order of the mappings by drawing the graph that was in the second pixel map graph in the first pixel map graph and drawing the graph that was in the first pixel map graph in the second pixel map graph.
	Q6: Is Output Image 2 still close to the Original Image? Would you have predicted this using mathematics?

Steps	Instructions (Continued)
	Q7: Describe the Output Image. How does it look compared to the image from Question 4?
Step 12:	Draw a map in the first pixel map graph that thresholds the original image with a threshold value of 128. Try to draw an inverse function in the second pixel map graph.
	Q8: Can a threshold function have an inverse function? Why or why not?
Step 13:	Draw a map in the first pixel map graph that looks like a capital U. Try to draw an inverse function in the second pixel map graph.
	Q9: Can this function have an inverse? Why or why not?
Step 14:	Stop the worksheet. Close the worksheet.

4.2.5 Threshold and Negation - Pixel Map Grayscale

Worksheet Description

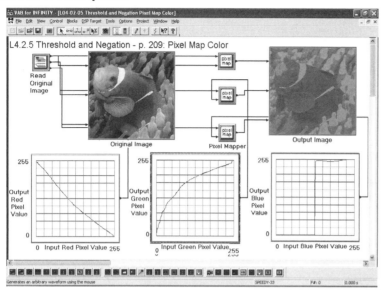

Figure 4.9 Threshold and Negation - Pixel Map Grayscale

Open the worksheet *L04-02-05 Threshold and Negation Pixel Map Color.Lst*. You should see a screen that looks like Figure 4.9 above except that the image displays will be blank and the pixel map graphs will have horizontal lines.

This worksheet is exactly like the worksheet for Lab 4.2.4 except that for a color image each of the three color components must have a pixel map graph and a Pixel Mapper block.

Laboratory Procedures

Steps	Instructions
Step 1:	Start the worksheet.
Step 2:	With the cursor, draw as straight a line as you can in the Red pixel map graph from the lower left corner to the upper right corner. Draw the same thing in the Green and Blue pixel map graphs.
	Q1: Describe the output image.
Step 3:	Draw pixel map graphs that brighten the blue component and darken the red component. Do not change the green component.

Steps	Instructions (Continued)
	Q2: Describe the output image.
	Q3: Draw the three pixel map graphs here.
Step 4:	Draw pixel map graphs that brighten the red component and darken the blue component. Do not change the green component.
	Q4: Describe the output image.
	Q5: Draw the three pixel map graphs here.
Step 5:	Draw pixel map graphs that create what you think is the most interesting output image.
	Q6: Describe the output image.
	Q7: Draw the three pixel map graphs here.
Step 6:	Stop the worksheet. Close the worksheet.

4.2.6   Threshold and Negation - Pixel Map Color Camera

Worksheet Description

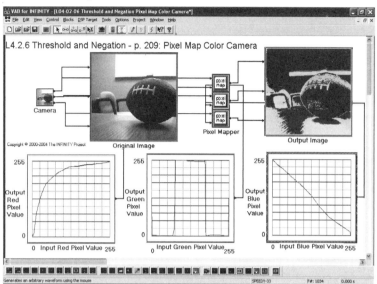

Figure 4.10 Threshold and Negation - Pixel Map Color Camera

This worksheet is exactly like the previous worksheet except that a camera block supplies the color input image instead of a file read block. Use this lab to create interesting images from the world around you.

Overview Questions

A: If you threshold an image at level v and then take the negative of it, write an expression for the relationship between the output pixel values and the input pixel values.

B: How could you use mapping on a color image to make it look as if it were taken at sunset instead of in the middle of the day?

C: How can adjusting the threshold be used to find out what value a pixel has.

Summary

Using a table or a graph to "map" input pixel values to output pixel values is a powerful digital darkroom tool. We can easily create maps for traditional film based darkroom effects such as thresholding, creating a negative, brightening, or contrast adjustment. The digital mapping method is very flexible, and it can create many other effects that would be impossible for film based processing.

4.3   Adding and Subtracting Images

Lab Objectives

In this demonstration laboratory we will look at the sum and difference of two images taken in the same place at different times, and we will explore what blocks we can use to do arithmetic on images.

Textbook Reading

*   This lab appears on page **216** of the *Engineering Our Digital Future* textbook.
*   Prerequisite textbook reading before completing this lab: pp. **213-216**.

Engineering Designs and Resources

Worksheets used in this lab:

*   **L04-03-01 Adding and Subtracting Images:** Students observe effects of adding and sub-tracting two color images.

Worksheet Description

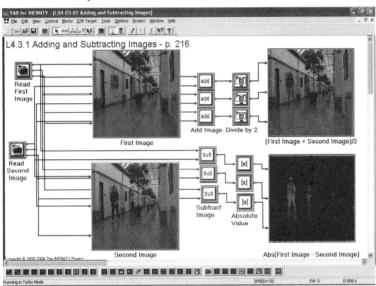

Figure 4.11 Threshold and Negation - Pixel Map Grayscale

Open the worksheet *L04-03-01 Adding and Subtracting Images.Lst.* You should see a screen that looks like Figure 4.11 above except that the image and numeric displays will be blank.

On the left side of the worksheet two images are read using two Bitmap Read blocks, and each is displayed. Although these two images could be any two images, in this laboratory we will use two images of the same scene taken at different times.

The two images are added using an Add Image block for each color component. Following each Add Image block is a Multiply by Constant block. This block multiplies each sum by 0.5 so the output will still be in the normal 8-bit range of 0 to 255. The image sum after scaling by 0.5 is shown in the display in the upper right. The second image is subtracted from the first image using a Subtract Image block for each color component. The output of each Subtract Image block goes through an Image Absolute Value block to make any negative differences positive and keep the result in the normal 8-bit range of 0 to 255. The lower right display shows the difference image.

Laboratory Procedures

Steps	Instructions
Step 1:	Start the worksheet.
Step 2:	Left click each input wire to the three Add Image blocks to select them one at a time and see how they are connected. Do the same for the Subtract Image blocks.
	Q1: Symbolically, write an expression for the output of each Add Image and Subtract Image block using input names like First Image Red or Second Image Blue.
	Q2: In the sum image, why do the people in the hallway have a "ghostly" look compared to the second image? For the person on the left side of the image, why is the clothing a lighter color than it is in the original image?
	Q3: In the difference image, list the objects that you see against the dark background. -How many people do you see? Do you notice more that you did when you first viewed the second image? -What else do you see in the difference image? Why?
	Q4: Why does the person on the left side of the difference image appear to be wearing white pants with high dark socks and a red shirt? You will need to examine both image 1 and image 2 to answer this.
Step 3:	Use the cursor to find the values of ten points in the apparently completely black areas of the image by clicking them, and record the RGB values.

Steps	Instructions (Continued)
	Q5: What is the largest value you recorded? Why aren't these values all zero?
Step 4:	Right click one the **Bitmap Read** blocks.
	Q6: What parameters does it have and how are they set?
Step 5:	Right click one of the **Image Add** blocks to see its parameters.
	Q7: What parameters does it have and how are they set?
Step 6:	Right click one of the **Multiply by Constant** blocks to see its parameters.
	Q8: What parameters does it have and how are they set?
	Q9: What would happen if the Multiply by Constant blocks had not been used? Test your answer by right clicking each block and changing the constant from 0.5 to 1.0.
Step 7:	Right click one of the **Image Subtract** blocks to see its parameters.

Steps	Instructions (Continued)
	Q10: What parameters does it have and how are they set?
Step 8:	Right click one of the **Image Absolute Value** blocks to see its parameters.
	Q11: What parameters does it have and how are they set?
	Q12: What would happen if the Image Absolute Value blocks had not been used? If they had not been used, would the display of (Image 1 - Image 2) be different from the display of (Image 2 - Image 1)?
Step 9:	Stop the worksheet. Close the worksheet.

Further Exploration - Find the difference between two images taken at different times from a camera.

Steps	Instructions
Step 1:	Open the worksheet *L04-03-01 Adding and Subtracting Images.Lst*.
Step 2:	Add a camera and an image delay to the worksheet. From the drop-down menu of Blocks, click **Select Blocks**. Choose the Image Processing Library. • Select **Group List: Frame grabbers** and click **Standard Video Input Device** in the Function List. • Click **Add to Worksheet** • Select **Group List: General Functions** and click **Image Delay** in the Function List. • Click **Add to Worksheet** three times. • Click **Close** You should see the four new blocks on your worksheet.

Steps	Instructions (Continued)
Step 3:	Move the new blocks.
	Left click the **Camera** block and drag it to a position close to the Read First Image block.
	Also move all three image delay blocks to positions near the Read Second Image block.
Step 4:	Right click the **Standard Video Input Device** label and change it to "Camera."
	Right click the labels for the **Image Delay** blocks and change them to "Red Delay", "Green Delay" and "Blue Delay."
Step 5:	Make a delayed image.
	Click the **Connect** icon. Connect each of the camera outputs to the input of the corresponding Image Delay block.
Step 6:	Replace the Read First Image block with the Camera block.
	• Click the **Delete** icon. (It looks like a pair of scissors.)
	• Left click the **Read First Image** block to delete it and all the wires connected to it.
	• Click the **Connect** icon
	• Connect the camera outputs to the First Image display and to the Add Image blocks and Subtract Image blocks in the same way that the Read First Image block was connected. (It might be helpful to refer the answer to Question 1 in the basic laboratory procedure.)
Step 7:	Replace the Read Second Image block with the three delayed image components.
	• Click the **Delete** icon. (It looks like a pair of scissors.)
	• Left click the **Read Second Image** block to delete it and all the wires connected to it.
	• Click the **Connect** icon
	• Connect the Image Delay block outputs to the Second Image display and to the Add Image Blocks and Subtract Image Blocks in the same way that the Read Second Image block was connected. (It might be helpful to refer the answer to Question 1 in the basic laboratory procedure.)

Steps	Instructions (Continued)
Step 8:	We are almost ready to start our worksheet, but first we have to make a few changes because our data are coming from the camera instead of a file. • Right click each of the three **Image Subtract** blocks and change the precision from Auto to Short. • Right click each of the **Multiply by Constant** blocks and change the constant from 0.5 to 1.0. 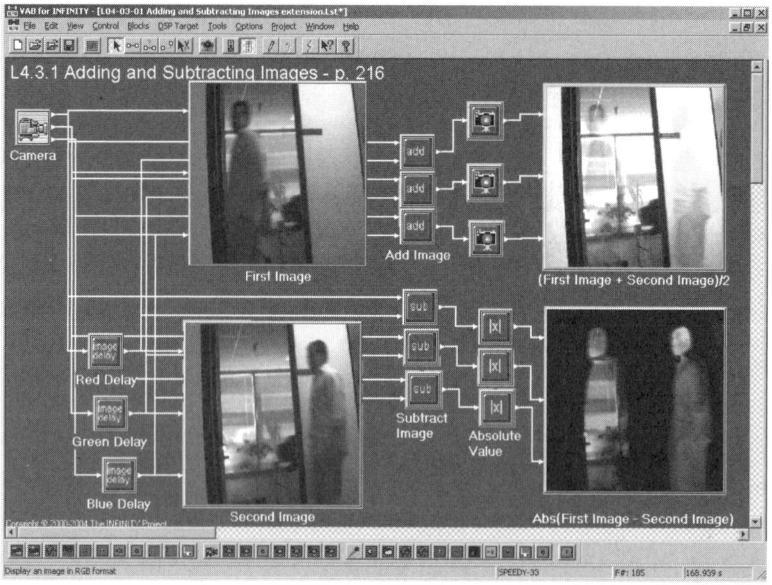 **Figure 4.12** Sum and difference of delayed images from a camera Your worksheet should now look similar to Figure 4.12.
Step 9:	Start your worksheet. Place the camera on a solid surface and point it at a region of the room where there are many distinct objects but nothing is moving.
	Q1: What do the sum and difference images look like? Why?
	Q2: Could this worksheet be used to detect vibrations of the surface holding the camera?

Steps	Instructions (Continued)
	Q3: Could this worksheet be used to alert you to the motion of objects within the camera's field of view?
Step 10:	Hold the camera in your hand and point it at a region of the room where there are many distinct objects but nothing is moving.
	Q4: Can you hold the camera still enough so that the difference image is always dark? Based on this, what advantage would a tripod or other image stabilization methods provide for long exposure photographs?
Step 11:	Stop the worksheet. Close the worksheet.

Overview Questions

A: When a difference of two images is computed, compare the results of setting all negative values to zero to the results of taking the absolute value of the difference.

B: Why does adding images give the appearance of transparency to objects that are in only one image?

Summary

Adding and subtracting images is useful for many imaging applications. Subtracting images taken at different times can detect objects that have moved. Adding low quality images taken at different times can smooth out defects. Adding images that are similar gives a transparent appearance to the parts that are different. In all cases consideration must be given to how results will be handled if they are not between 0 and 255.

4.4   Adding and Subtracting Shifted Images

Lab Objectives

For robot vision and many other imaging applications we need a way to figure out what is in an image. In this laboratory we will explore what happens when we add and subtract an image and a shifted version of the same image. We will discover that the sum and difference depend both on what is in the image and on how much we shift it. Because of this, combining shifted images can be used to automatically understand images.

Textbook Reading

*   This lab appears on page **223** of the *Engineering Our Digital Future* textbook.
*   Prerequisite textbook reading before completing this lab: pp. **218-224**.

Engineering Designs and Resources

Worksheets used in this lab:

*   **L04-04-01 Adding and Subtracting Shifted Images.Lst:** Students observe effects of adding and subtracting an image and a shifted image.
*   **L04-04-02 Adding and Subtracting Shifted Images Camer.Lst:** Students observe effects of adding and subtracting an image and a shifted image from a camera.

4.4.1   Adding and Subtracting Shifted Images

Worksheet Description

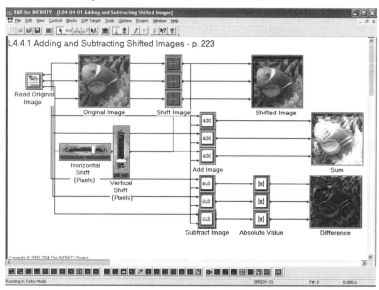

Figure 4.13 Adding and Subtracting Shifted Images

Open the worksheet *L04-04-01 Adding and Subtracting Shifted Images.Lst*. You should see a screen that looks like Figure 4.13 except that the image displays will be blank.

At the left edge of the worksheet the "Read Original Image" block reads a bitmap image file. The three color component outputs of the bitmap read block are connected to the corresponding three inputs of the display labeled "Original Image." Three other displays show a shifted image and the sum and difference of the original and shifted images.

A shifted image is created using a shift image block for each color component. The amount of horizontal and vertical shift for all three image shift blocks is controlled by two sliders. For each color component, the input of the shift image block is connected to the output of the bitmap read block. The output of the shift image block is connected to the input of the Shifted Image display.

The sum and difference of the original and shifted images are created and displayed as they were in Lab 4.3. The only difference is that here the image sum is not divided by 2.

Laboratory Procedures

Steps	Instructions
Step 1:	Start the worksheet.
Step 2:	Set both shift values to 0.
	Q1: What should the image sum be? Explain the appearance of the image sum in terms of the original image.
	Q2: What should the image difference be? How do you think the display should look? How does the display look?
Step 3:	Slowly change the horizontal shift value from 0 to 50 and observe both the sum and difference images as the shift value increases.
	Q3: What values are shifted into the left edge of the shifted image as it moves to the right?

Steps	Instructions (Continued)
	Q4: What do the sum and difference images look like in the 50 pixel wide band at the left? Why?
Step 4:	Slowly change the horizontal shift value from 0 to -50 and observe both the sum and difference images as the shift value increases.
	Q5: How are results different for a -50 shift and a + 50 shift?
Step 5:	Return to a horizontal shift of +10
	Q6: How does the sum image look compared to a shift of zero? Does it look like a "double exposure" to you? Why would you expect this?
	Q7: How does the difference image look compared to a shift of zero?
	Q8: Which edges are easy to see in the difference image? Which edges are not visible in the difference image?
Step 6:	Set the horizontal shift to 0. Slowly shift up and down from -50 to +50 using the vertical shift slider.

Steps	Instructions (Continued)
	Q9: How do the sum and difference images compare to the images observed earlier with a shift left and right from -50 to 50?
Step 7:	Return to a vertical shift of +10.
	Q10: What edges can you see in the subtracted image? How are these different from edges using a horizontal shift of +10. Why are they different?
Step 8:	Compare the difference image for a vertical shift of -34 and a vertical shift of +34.
	Q11: What do you see when you look at the stripes on the fish for the two shifts? Why are they different? (Be careful. Note that each image subtract output goes through an absolute value block before it is displayed.)
Step 9:	Set the horizontal shift to +20 and the vertical shift to -30.
	Q12: What do you see in the difference image? What do you see in the sum image?
Step 10:	Stop the worksheet. Close the worksheet.

Optional Quantitative Investigation: Alternative sum and difference images
The sum and difference images we observed earlier in this laboratory are not the only possibilities for these images. Both the image adders and the image subtractors will produce different

images if they deal with out of range values in a different way. These images are also interesting and may be useful in other applications. We will briefly explore some of these alternatives.

Steps	Instructions
Step 1:	Open the worksheet *L04-04-01 Adding and Subtracting Shifted Images* and start it. Set both sliders to a value of zero.
	Q1: How do the colors in the sum image compare to the colors in the original image?
Step 2:	Right click one of the **image add** blocks. You will see that the option "clip to max" is selected with a "yes." If two image pixels with values between 0 and 255 are added, the resulting pixel will have a value between 0 and 510. With "clip to max" selected, all resulting pixels greater than the maximum value of 255 will be set to 255. This is also often called "saturation." Click **cancel** to leave the options unchanged for the moment.
	Q2: Does this clipping explain your observations of the sum image in Question 1? What happens to dark areas of the original image? What happens to bright areas of the original image?
	Q3: Find three good examples of dark colors in the original image. Place the cursor over them and click to see what the RGB values are. Find the RGB values in corresponding areas of the sum image. (You will not be able to find exactly the same pixels in both images, so choose an area where there is not much color variation.) How do the values compare?
	Q4: Find three good examples of bright colors in the original image and repeat Question 3. Does "clip to max" explain the values you have read?

Steps	Instructions (Continued)
Step 3:	Turn off the "clip to max" option for the image add blocks. • Right click one of the **image add** blocks as we did in step 2. • Select **No** for **clip to max.** • Then click **OK**. • Do the same for the other two image add blocks. Now observe the sum image.
	Q5: How does the new sum image compare to the sum image observed in Question 1?
	Q6: Find areas of the original image that are the same color, but brighter, in the sum image. Find areas of the original image where the color is very different in the sum image. Repeat Question 3 for these areas. What happens when "clip to max" is turned off? How does this explain the color changes?
Step 4:	Compare the difference image for a vertical shift of -34 and a vertical shift of +34.
Step 5:	Change the parameters for the image subtract blocks. If two image pixels with values between 0 and 255 are subtracted, the resulting pixel will have a value between -255 and 255. With "clip to max" selected, all resulting pixels with a value less than 0 will be set to 0. • Right click one of the **image subtract** blocks. • Turn off "clip to max" by selecting **No**. • Set the precision to short. • Then click **OK**. • Do the same for the other two image subtract blocks.
Step 6:	Repeat Step 4.

Steps	Instructions (Continued)
	Q7: Are the difference images the same now for both shifts? (Ignore the filled regions at the top and bottom of the difference image for this question.) How are the images different from what you observed in Step 4? What effect does "clip to max" have on image subtraction?
Step 7:	Stop the worksheet. Close the worksheet.

4.4.2 Adding and Subtracting Shifted Images Camera

Worksheet Description

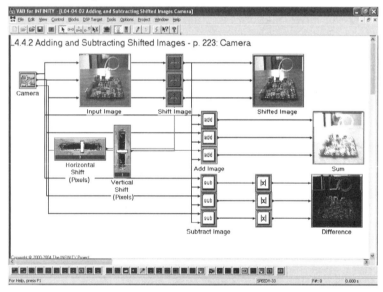

Figure 4.14 Adding and Subtracting Shifted Images - Camera

This worksheet is exactly like the previous worksheet except that the file read block is replaced with a camera block.

Laboratory Procedures

Aim the camera at a variety of different scenes and see if you observe the same effects you saw in the previous laboratory. For example, find a striped object and try to make the shift match the width of the stripes.

Overview Questions

A: Why do the sum and difference of shifted images depend on the size and shape of objects in the image?

B: If an image and a shifted version of the image are added and the sum image is a constant value, does that means that the original image must be a constant value? Why or why not? Give some examples.

Summary

The effects of adding and subtracting shifted images depends very much on the image content. Areas where the intensity is slowly varying have dramatically different results than areas with fine detail or sharp edges. That means that we should be able to use combinations of shifted images to detect different kinds of images and figure out what kinds of objects are shown in an image.

4.5 Sharpening Images

Lab Objectives

People almost always prefer a sharp crisp image to an image with soft edges. Edges can be soft-ened or blurred when some objects in an image are moving or out of focus. In this laboratory you will see how digital darkroom sharpening methods can make a blurred image look much better. You will learn how the sharpening method works and see what happens when images are over sharpened.

Textbook Reading

*   This lab appears on page **231** of the *Engineering Our Digital Future* textbook.
*   Prerequisite textbook reading before completing this lab: pp. **224-230**.

Engineering Designs and Resources

Worksheet used in this lab:

*   **L04-05-01 Sharpening Images.Lst:** Students examine the effects of sharpening filters on digital images.

Worksheet Description

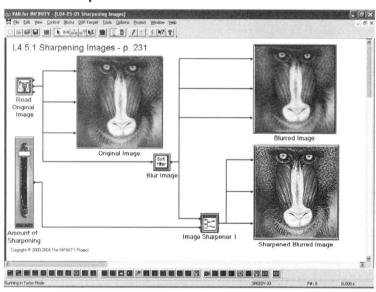

Figure 4.15 Sharpening Images

Open the worksheet *L04-05-01 Sharpening Images.Lst.* You should see a screen that looks like Figure 4.15 above except that the image displays will be blank.

At the left edge of the worksheet the "Read Original Image" block reads a bitmap image file. As we have done before, we will use the green color component output from this block to represent a grayscale image. The output of the bitmap read block is connected to all three inputs of the display labeled "Original Image."

The input image is blurred by a 5x5 filter block which has parameters set to create a small but noticeable blur. The image input of the 5x5 filter block comes from the image output of the bitmap read block. The image output of the filter block is connected to all three inputs of the blurred image display. The function of the filter blocks is to add 25 shifted versions of the input image together to make the output image using the following filter parameters (This is similar to Lab 4.4 except that 25 images may be added instead of only 2.). Each parameter is a scale factor that multiples a shifted image. The row of the parameter array indicates the amount of vertical shift and the column indicates the amount of horizontal shift.

Vertical Shift	Horizontal Shift				
	-2	-1	0	1	2
-2	0	0	0	0	0
-1	0	0.1	0.1	0.1	0
0	0	0.1	0.2	0.1	0
1	0	0.1	0.1	0.1	0
2	0	0	0	0	0

From this table we can see that only 9 of the 25 possible shifted images are added because 16 of the filter parameters are zero.

The image sharpening block will make the blurred image look sharper and reduce the appearance of blur. It does this by increasing the difference between nearby pixels values.

In areas of the image that have a constant intensity, there is no change, so the whole image appears unchanged except that it has brighter edges. There are two inputs to the image sharpening block. One is the blurred image, and the other is a value from a slider which controls the amount of sharpening. The image output of the image sharpener block is connected to all three inputs of the Sharpened Blurred Image display.

Laboratory Procedures

Steps	Instructions
Step 1:	Start the worksheet, and adjust the amount of sharpening to 0.
Step 2:	Observe the blurred image and the sharpened image.
	Q1: How does the blurred image differ from the original image?

Steps	Instructions (Continued)
	Q2: Does the sharpened image look the same as the blurred image when the amount of sharpening is 0?
Step 3:	Right click the 5x5 filter block and verify that its parameters are the same as the ones given in the worksheet description. You should see a pop up window like the one shown here. Then click the **cancel** button.

General Filter (5x5) Parameters

A1: 0.0	B1: 0.0	C1: 0.0	D1: 0.0	E1: 0.0
A2: 0.0	B2: 0.1	C2: 0.1	D2: 0.1	E2: 0.0
A3: 0.0	B3: 0.1	C3:* 0.2	D3: 0.1	E3: 0.0
A4: 0.0	B4: 0.1	C4: 0.1	D4: 0.1	E4: 0.0
A5: 0.0	B5: 0.0	C5: 0.0	D5: 0.0	E5: 0.0

* C3 represents the weighting coefficient for the center pixel.

Precision: Auto

OK   Cancel   Help

Steps	Instructions (Continued)
	Q3: With the filter parameters set as they are in the image above, what is the maximum value that the sum of shifted images can have? (Hint: Since all the shifted images are multiplied by positive values, just assume that all the shifted images have the maximum value.)
Step 4:	Using the slider, increase the amount of sharpening to 0.2 and compare the sharpened image to the original image and the blurred image.

Steps	Instructions (Continued)
	Q4: Where do you see differences? Look for the finest details such as the small whiskers on the baboon.
Step 5:	Increase the amount of sharpening to 0.33 and compare the sharpened image to the original image and the blurred image in the specific areas listed.
	Q5: The nose The fur The whiskers The eyes
	Q6: Does the sharpened image look more like the original image than it did with a lower amount of sharpening?
Step 6:	Increase the amount of sharpening to the maximum of 1.0 and compare the sharpened image to the original image. Pay special attention to the areas of fine detailed patterns such as the fur.
	Q7: Does the sharpened image look realistic? Why or why not? Does it look "sharp"?
Step 7:	Adjust the amount of sharpening to the value that makes the sharpened image look most like the original image to you.

Steps	Instructions (Continued)
	Q8: Record this value and compare it to the values selected by other students.
Step 8:	Right click the slider and set the top value to 2 instead of 1. Then click **OK**. Set the amount of sharpening to 2 and compare the sharpened image to the original image.
	Q9: Describe the sharpened image now.
Step 9:	Now we will use the image sharpener on a different image. With the right mouse button, click the **Read Original Image** block. Click the **Browse** button and then navigate to Program Files\Hyperception\VABINF\Images\BWimages and select *graybands640x480.bmp* in this directory. Click **Open** in the browser's Open window, and then click **OK** in the Bitmap Read Parameters Window.
	Q10: How is the blurred image different from the original image?
Step 10:	Set the amount of sharpening to 0 and slowly increase it from 0 to 0.5 while observing the sharpened image.
	Q11: What changes do you see in the sharpened image? How is the lowest band different from the other bands?
Step 11:	Continue to increase the amount of sharpening.
Step 12:	Stop the worksheet. Close the worksheet.

Further exploration: What effect does changing the amount of blur have on the amount of sharpening that looks best?

Steps	Instructions
Step 1:	Open the worksheet *L04-05-01 Sharpening Images.Lst.* Start the worksheet. Look at the blurred image compared to the original image.

Steps	Instructions (Continued)
Step 2:	Increase the blur. Right click the 5x5 filter and change the filter parameters to be: 0.1 0 0.1 0 0.1 0 0 0 0 0 0.1 0 0.2 0 0.1 0 0 0 0 0 0.1 0 0.1 0 0.1 Click **OK**.
	Q1: How does this blurred image differ from the original image? Does it seem more blurred than it was before we made the change in scale factors?
Step 3:	Adjust the amount of sharpening to the value that makes the sharpened image look most like the original image to you.
	Q2: Record this value and compare it to the values selected by other students. How does this value compare to the value you recorded in Step 7 of the earlier laboratory procedure?
Step 4:	Reduce the blur. Right click the 5x5 filter and change the filter parameters to be: 0 0 0 0 0 0 0 0.1 0 0 0 0.1 0.6 0.1 0 0 0 0.1 0 0 0 0 0 0 0 Click **OK**.
	Q3: How does this blurred image differ from the original image? Can you find the part of the image where the blur is most noticeable?

Steps	Instructions (Continued)
Step 5:	Adjust the amount of sharpening to the value that makes the sharpened image look most like the original image to you.
	Q4: Record this value and compare it to the values selected by other students. How does this value compare to the value you recorded in Step 7 of the earlier laboratory procedure? How does it compare to the value you found in Step 3 above?
Step 6:	Eliminate the blur. Right click the 5x5 filter and change the scale factors to be: 0 0 0 0 0 0 0 0 0 0 0 0 1.0 0 0 0 0 0 0 0 0 0 0 0 0 Click **OK**.
Step 7:	Adjust the amount of sharpening to the value that makes the sharpened image look the best to you.
	Q5: Record this value and compare it to the values selected by other students. Why might this value be larger than 0?
	Q6: Could you have tested the case where there was no blur by rewiring the worksheet instead of changing the parameters as we did in Step 6? How would you have done that?
Step 8:	Stop the worksheet. Close the worksheet.

Further exploration: How does the sharpening filter work?

Steps	Instructions
Step 1:	Open the worksheet *L04-05-01 Sharpening Images.Lst*.
Step 2:	Start the worksheet.
Step 3:	Right click the slider. Note that slider output is Ax + b and A is -1. Set slider value to 0.204
Step 4:	The image sharpener block is a hierarchy block. It appears as one block in our design, but it is made from several VAB blocks. Double click the image sharpener block to see what blocks are inside it. You should see a screen that looks like Figure 4.16 below. 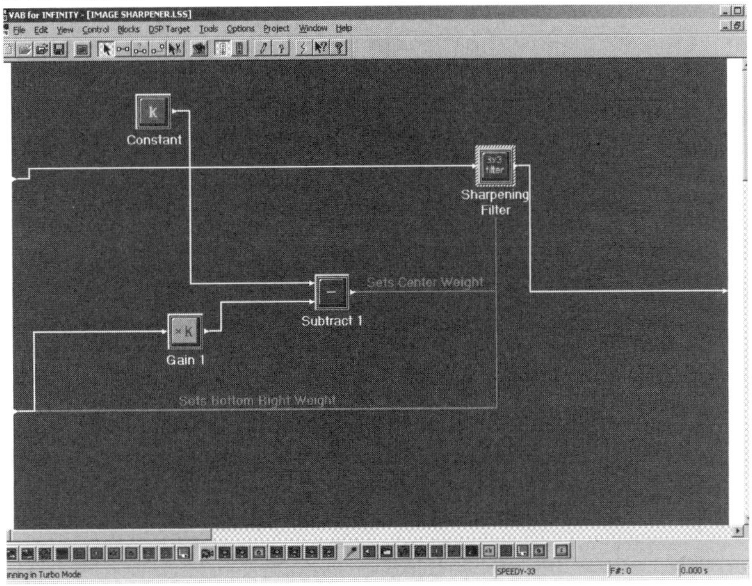 **Figure 4.16** Image sharpener block • The upper input at the left edge of the screen is the blurred image input. • The lower input is the slider value multiplied by -1.
Step 5:	Explore the blocks: • Right click the gain 1 block and verify that it has a value of 8. • Right click the constant block and verify that it has a value of 1.

Steps	Instructions (Continued)
	Q1: If the slider is set to a value of C, then the lower input to the sharpening filter block shown above has a value of -C. -Write an expression for the green parameter line "sets bottom right weight" in terms of C. -Write an expression for the green parameter line "sets center weight" in terms of C.
	Q2: Evaluate the two expressions in Q1 for our slider value of C = 0.204.
Step 6:	Right click the 3x3 filter block. • Verify that all values except the center value are close to -C where C = 0.204. Find the difference between the expected value of -C and the value you see in the filter. • Record the center value. • Use the method of Question 1 to compute the center value of the filter for the value of C that you actually observed, and verify that it is the value you just recorded
	Q3: If you round off the value you observed for C, would it be the same as the slider value? What would cause the slider value and the value you see in the filter to be different?

Steps	Instructions (Continued)
	Q4: If C = 0.2, what would the output of the sharpening filter be for the following input images? -All pixels in the input are 100. -All pixels on the left side of the image are 90 and all pixels on the right side of the image are 110. -All pixels on the left side of the image are 80 and all pixels on the right side of the image are 120. -All pixels on the left side of the image are 50 and all pixels on the right side of the image are 150.
	Q5: Based on your answer to Question 3, explain how the sharpening filter block makes edges brighter but does not change areas of the image that are all the same value.
	Q6: Repeat Question 4 for C = 0.5. -All pixels in the input are 100. -All pixels on the left side of the image are 90 and all pixels on the right side of the image are 110. -All pixels on the left side of the image are 80 and all pixels on the right side of the image are 120. -All pixels on the left side of the image are 50 and all pixels on the right side of the image are 150.
	Q7: What effect would "clip to max" have on your results in Question 6?

Steps	Instructions (Continued)
Step 7:	Click the vertical blue arrow that is two icons to the right of the stop worksheet icon. This takes you back up to the worksheet.
Step 8:	Stop the worksheet. Close the worksheet.

Overview Questions

A: Does sharpening perfectly undo the image blur or is there always a difference between the sharpened blurred image and the original image?

B: When does sharpening improve images?

C: When does sharpening make images look worse?

D: How do we know when we have the right amount of sharpening?

Summary

Image sharpening improves the appearance of images that have soft edges. The right amount of sharpening brightens the edges in an image, but too much sharpening makes the image look worse. Sharpening often does not work well on images with a lot of fine detail.

4.6 Edge Detectors

Lab Objectives

When we describe an image to someone, we almost always start by identifying the objects we see in the image. For example, we might say the image shows a sailboat tied to a pier or books and a guitar on a chair. In robotic applications it is important for the robot's image processor to be able to locate objects automatically. Although it is very easy for us to find objects in an image, an automatic image processor must use several steps to do this. One of the most important is called edge detection. In this demonstration laboratory you will see the result of using several different edge detectors. You will also compare the edge detector to the sharpening filter of Lab 4.5 and the subtraction of shifted images in Lab 4.4.

Textbook Reading

- This lab appears on page **232** of the *Engineering our Digital Future* textbook.
- Prerequisite textbook reading before completing the lab: pp. **230-232**.

Engineering Designs and Resources

Worksheets used in this lab:

- **L04-06-01 Edge Detectors Grayscale Image.Lst:** Students observe three different edge detectors on a grayscale image.
- **L04-06-02 Edge Detectors Grayscale Camera.Lst:** Students observe three different edge detectors on a camera image.
- **L04-06-03 Edge Detectors Color Image.Lst:** Students observe three different edge detectors on a color image.

4.6.1 Edge Detectors - Grayscale

Worksheet Description

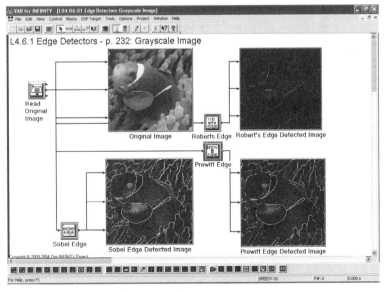

Figure 4.17 Edge Detectors - Grayscale Image

Open the worksheet *L04-06-01 Edge Detectors Grayscale.Lst.* You should see a screen that looks like Figure 4.17 above except that the image displays will be blank.

At the left edge of the worksheet the "Read Original Image" block reads a bitmap image file. As we have done before, we will use the green color component output from this block to represent a grayscale image. The output of the bitmap read block is connected to all three inputs of the display labeled "Original Image." The other three displays will show the outputs from three different types of edge detector blocks.

An edge detector finds parts of the image where there is a sharp change in intensity no matter what the edge angle is, so it must compare changes in both the horizontal and vertical directions. At each location in an image, an edge detector tells us how strong an edge is and what angle it has. In this laboratory the edge detector blocks compute only the strength of the edges, but not the angle. The edge strength will always be positive or zero, so we do not need to take the absolute value as we did with the difference images.

- The Sobel edge detector block is shown in the lower left. Its image input comes from the output of the bitmap read block, and its image output is connected to all three inputs of the display labeled Sobel Edge Detected Image. This block uses two of the 3x3 filter blocks we used in the previous laboratory. The values used in those blocks to find vertical and horizontal differences are:

1/4	0	-1/4	and	-1/4	-1/2	-1/4
1/2	0	-1/2		0	0	0
1/4	0	-1/4		1/4	1/2	1/4

- The Prewitt edge detector block, shown in the lower right, is very similar to the Sobel edge detector block. Its image input comes from the bitmap read block and its image output is connected to all three inputs of the display labeled Prewitt Edge Detected Image. This block also uses two 3x3 filter blocks. The values used in those blocks are:

1/3	0	-1/3	and	-1/3	-1/3	-1/3
1/3	0	-1/3		0	0	0
1/3	0	-1/3		1/3	1/3	1/3

- The Robert's edge detector block, shown in the upper right, uses diagonal differences instead of horizontal and vertical differences to find edges. Its image input comes from the output of the bitmap read block, and its image output is connected to all three inputs of the display labeled Robert's Edge Detected Image. This block also uses two 2x2 filter blocks. The values used in those blocks are:

1	0	and	0	-1
0	-1		1	0

There are two other edge detector blocks in the VAB library that may be substituted for the three blocks shown. These are the Isotropic Edge Detector Block and the Laplace Edge Detector Block.

Laboratory Procedures

Steps	Instructions
Step 1:	Start the worksheet.
Step 2:	Observe the three edge detected images.
	Q1: How are the Sobel and Prewitt Edge detected images different from the Robert's Edge Detected Image?
	Q2: Compare the three edge detectors on parts of the original image which have horizontal, vertical, and diagonal edges.
	Q3: Why are areas of slowly varying intensity in the image dark in the edge detected images? Look particularly at the area around the mouth of the fish and along its back.
	Q4: Find the brightest parts of the edge detected images and describe those areas in the original image.
Step 3:	Compare the edge detected images in this laboratory to the sharpened image output for this image in Lab 4.5.1.
	Q5: What differences do you see?

Steps	Instructions (Continued)
Step 4:	Compare the edge detected images in this laboratory to the differences of shifted images for this image in Lab 4.4.1
	Q6: What differences do you see?
Step 5:	Now we will use the edge detectors on a different image. With the right mouse button, click the **Read Original Image** block. Click the **Browse** button and then navigate to Program Files\Hyperception\VABINF\Images\BWimages and select *building256x256.bmp* from this directory. Click **Open** in the browser's Open window, and then click **OK** in the Bitmap Read Parameters Window.
	Q7: Repeat Question 4 for this image.
Step 6:	Try another image.
	Select the bitmap read block by clicking it.
	Navigate to Program Files\Hyperception\VABINF\Images\BWimages and select *StripePat1at256x256.bmp* in this directory.
	Q8: Repeat Question 4 for this image.
Step 7:	Try an image where we know what the differences are between nearby pixels values.
	Select the bitmap read block by clicking it.
	Navigate to Program Files\Hyperception\VABINF\Images\BWimages and select *graybands640x480.bmp* in this directory or a directory of test images.

Steps	Instructions (Continued)
	Q9: How is the brightness of the edge detected images related to the intensity differences between nearby pixels?
	Q10: How is the brightness of the edge detected images related to the average intensity values of the nearby pixels?
Step 8:	Stop the worksheet. Close the worksheet.

4.6.2   Edge Detectors - Grayscale Camera

Worksheet Description

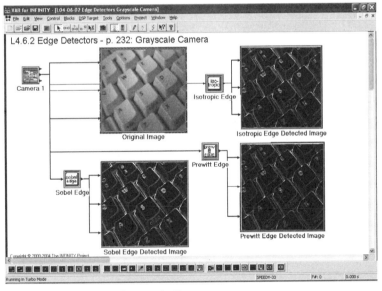

Figure 4.18 Edge Detectors - Grayscale Camera

Open the worksheet *L04-06-02 Edge Detectors Grayscale Camera*. You should see a screen that looks like Figure 4.18 above except that the image displays will be blank.

This worksheet is exactly the same as the previous worksheet except that the bitmap read block has been replaced with a Camera block. The Robert's Edge Detector block has also been replaced with the Isotropic Edge Detector Block.

Laboratory Procedure

Steps	Instructions
Step 1:	Start the worksheet
Step 2:	Aim the camera at the following objects and explain why the edge detected images appear as they do. The three edge detected images will all look very much alike, so use the Sobel Edge Detected Image for your answers.
	Q1: A page of a text book A pencil on a table A window and window frame (preferably with window blinds) A bowl of small objects such as uncooked rice Your shoes A sweater
	Q2: What happens when you move the camera closer to the objects or further away from the objects?
Step 3:	Stop the worksheet. Close the worksheet.

4.6.3 Edge Detectors - Color

Worksheet Description

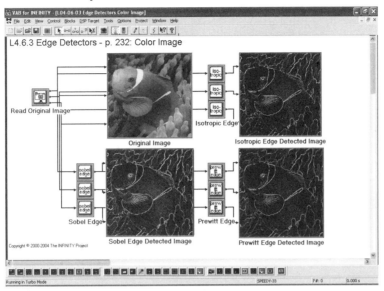

Figure 4.19 Edge Detector - Color Image

Open the worksheet *L04-06-03 Edge Detectors Color Image.Lst.* You should see a screen that looks like Figure 4.19 above except that the image displays will be blank.

This worksheet is exactly the same as the previous worksheet for Lab 4.6.1 except that color images are used. Now each edge detected image will need three edge detector blocks, one for each of the three color components. In addition, the Robert's Edge Detector block has also been replaced with the Isotropic Edge Detector Block for this laboratory.

Laboratory Procedures

Steps	Instructions
Step 1:	Start the worksheet.
Step 2:	Observe the three edge detected images. They will all look very much alike, so use the Sobel Edge Detected Image when answering the questions below.
	Q1: Find edges in the edge detected image that look white. In that area of the original image, what are the two colors on each side of the edge?

Steps	Instructions (Continued)
	Q2: Find edges in the edge detected image that look blue. In that area of the original image, what are the two colors on each side of the edge? Why are the edges blue? (Hint: What are the red, green, and blue color components of yellow and orange?)
	Q3: Find edges in the edge detected image that look bright orange or yellow. In that area of the original image, what are the two colors on each side of the edge? Why are the edges orange?
Step 3:	Use the edge detectors on a other color images such as *building256x256.bmp*.
	Q4: When the image has a colored object and a black background, what color is the edge? Why?
	Q5: When the image has a colored object and a white background, what color is the edge? Why?
Step 4:	Stop the worksheet. Close the worksheet.

Further Exploration

You can easily make a worksheet that will show edge detected images from a color camera. How would you modify Lab 4.6.2 to do this? How would you modify Lab 4.6.3 to do this? Which would be easier?

Overview Questions

A: How is edge detection different from image sharpening?

B: How would brightening an image change the edge detected image? Why?

C: How would sharpening an image change the edge detected image? Why?

Summary

Image edge detection blocks find edges at all angles and tell us how strong the edges are. The strong edges are used to find the boundaries of objects in images. For many imaging applications this is one of the first steps in figuring out what objects are in an image.

4.7 Averaging and Median Smoothing Filters

Lab Objectives

In this laboratory you will compare two different methods that smooth out imperfections or noise in images. The tricky part of smoothing out imperfections is that sometimes small but important parts of the image look just like imperfections. Often smoothing also makes sharp edges of objects look blurry. Experiments with different amounts of smoothing on different images will demonstrate the trade-off between smoothing an image to improve its appearance and losing information in the image.

Textbook Reading

- This lab appears on page **236** of the Engineering our Digital Future textbook.
- Prerequisite textbook reading before completing this lab: pp. **231-237**.

Engineering Designs and Resources

Worksheets used in this lab:

- **L04-07-01 Averaging and Median Smoothing Filters Image.Lst:** Students adjust the amount of smoothing applied to a grayscale image.
- **L04-07-02 Averaging and Median Smoothing Filters Camera.Lst:** Students adjust the amount of smoothing applied to a camera image.
- **L04-07-03 Averaging and Median Smoothing Filters Noise.Lst:** Students apply smoothing filters to improve an image with added noise.

4.7.1 Averaging and Median Smoothing Filters - Image

Worksheet Description

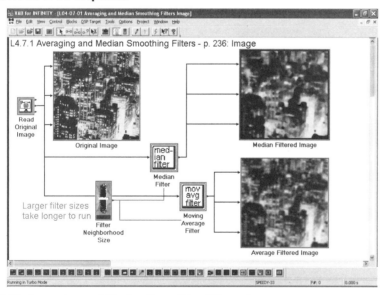

Figure 4.20 Averaging and Median Smoothing Filters - Image

Open the worksheet *L04-07-01 Average and Median Smoothing Filters Image.Lst*. You should see a screen that looks like Figure 4.20 above except that the image and numeric displays will be blank.

The Read Original Image block reads an image file into the worksheet. The output of this block is connected to all three inputs of the Original Image display and also to the inputs of two different smoothing blocks. Both blocks compute a smoothed output for each input pixel using a square n pixel by n pixel array. The value of n is controlled by the slider. The moving average filter adds all the pixel values in each square array and divides by the number of pixels to compute an average output. The median filter sorts all the pixels in each square array in increasing order and selects the center value. When the value of n increases above 5, this sorting can take a long time.

Laboratory Procedures

Steps	Instructions
Step 1:	Start the worksheet. Set the Filter Neighborhood Size slider to its minimum value of 3.
	Q1: Compare both images to the original. Look closely at the loss of detail in the sharp corners of the windows and on thin lines.
	Q2: What would you expect to see if the slider had set the neighborhood size to 1?
Step 2:	Increase the size of the neighborhood to 5.
	Q3: Now how do the smoothed images compare to the original? Find areas in the image where the median filtered image looks better than the averaged image. Find areas in the image where the averaged image looks better than the median filtered image.
Step 3:	Increase the size of the neighborhood to 7.

Steps	Instructions (Continued)
	Q4: Which filter does a better job of keeping the window details for the building in the lower left corner of the image?
	Q5: Which filter does a better job of preserving the sharp boundaries between large light and dark areas?
	Q6: Would you be able to identify the objects in the in the smoothed image if you did not have the original image for reference?
	Q7: Stand several feet away from the screen. Do the smoothed images and the original image looks more alike as you increase you distance from the screen? What would cause this?
Step 4:	Change the size of the neighborhood back to 3.
Step 5:	Compare these results to results from another image.
	Right click the **Read Original Image** block.
	Click the **Browse** button and then navigate to Program Files\Hyperception\VAB-INF\Images\BWimages and select either *BABOON256x256.bmp.*
	Click **Open** and then click **OK**.

Steps	Instructions (Continued)
	Q8: Do you see any difference between the smoothed images and the original image for the neighborhood size of 3? Look closely at: -The eyes -The fur around the face -The whiskers
Step 6:	Increase the size of the neighborhood to 7.
	Q9: Which smoothed image does a better job of keeping the whiskers? Which does a better job of keeping a sharp boundary between the light and dark areas of the nose?
	Q10: Would you be able to identify the objects in the in the smoothed image if you did not have the original image for reference? Why is this answer different from the answer to Question 6?
Step 7:	Increase the size of the neighborhood to 11. Be patient in waiting for the results to appear on the screen.
	Q11: Now, would you be able to identify the objects in the in the smoothed image if you did not have the original image for reference?

Steps	Instructions (Continued)
	Q12: What has happened to the baboon's eyes in the image? Why?
	Q13: Explain how the two smoothed images are different with respect to: -Retaining the detail of the original image -Retaining sharp boundaries between large dark and light areas
Step 8:	Set the size of the neighborhood to 3.
Step 9:	Stop the worksheet. Close the worksheet.

4.7.2   Averaging and Median Smoothing Filters - Camera

Worksheet Description

You may also find it interesting to see the effects of smoothing on images from the camera by opening worksheet L04-07-02 Averaging and Median Smoothing Filters - Camera. This worksheet is exactly like the previous worksheet in Lab 4.7.1 except that the image file reading block is replaced with a camera block.

4.7.3   Averaging and Median Smoothing Filters - Noise

Worksheet Description

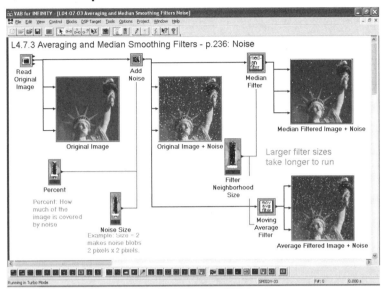

Figure 4.21 Averaging and Median Smoothing Filters - Noise

Open the worksheet *L04-07-03 Average and Median Smoothing Filters Noise.Lst*. You should see a screen that looks like Figure 4.21 above except that the image displays will be blank.

This worksheet allows us to make imperfections in our image and then see how well the smoothing filters reduce the appearance of the imperfections. Imperfections in images can come from aging photographs, or low light images, or images sent through poor communications systems.

This worksheet is based on the worksheet from Lab 4.7.1, which uses median and averaging filters to smooth an image. But the image that is smoothed in this laboratory has added imperfections.

The output of the Read Original Image block is connected to the input of an Add Noise block. This block changes some percentage of the pixels to a value set by "noise amplitude" in the block's parameters. The size and percentage of pixels that are changed are controlled by sliders. Because the output of the Add Noise block is the original image with these added imperfections, it is often called a noisy image. It is the noisy image that is connected to the inputs of the two smoothing blocks.

Laboratory Procedures

Steps	Instructions
Step 1:	Start the worksheet.
Step 2:	Set the noise percentage to 4%, the noise size to 1, and the neighborhood size to 3. Observe the images as the noise pattern that disturbs the image continually changes.

Steps	Instructions (Continued)
	Q1: Describe the noisy image compared to the original image. Right click the **Add Noise** block and find the value of the noise amplitude. How does this help explain the appearance of the noisy image?
	Q2: Describe the Average Filtered (Image + Noise) display with respect to the original image and the noisy image. What effect does the averaging filter have on the noise? How is the noise in this image different from the noise in the noisy image?
	Q3: Describe the Median Filtered (Image + Noise) display with respect to the original image and the noisy image. What effect does the median filter have on the noise? Do any noise pixels get through the median filter? Pay close attention to edges.
	Q4: How does the median filter completely remove most of the noise? Why would a median filter be better at removing noise in the open sky than at the edges of the statue?
Step 3:	Increase the percentage of the noise to 10%.
	Q5: How does this affect the noisy image, the averaged image, and the median filtered image? Why does more noise get through the median filter in the sky?

Steps	Instructions (Continued)
Step 4:	Increase the filter neighborhood size to 5.
	Q6: How does this change affect: -The median filtered image? -The features on the statue in the median filtered image? -The averaged image?
	Q7: How does this change affect the rate at which the noise is changing? What do you think this tells us about the relative amount of time it takes to compute a filtered image when the neighborhood size is increased?
Step 5:	Set the noise percentage to 4%, the noise size to 3, and the neighborhood size to 3.
	Q8: Describe the noisy image relative to the original image. How does increasing the noise size affect the appearance of the noisy image?
	Q9: Compare the averaged image to the median filtered image for this neighborhood size of 3. Is either filter doing a very good job of reducing the noise?
Step 6:	Increase the filter neighborhood size to 5.
	Q10: What happens to the noise in each of the smoothed images? Why?

Steps	Instructions (Continued)
Step 7:	Increase the filter neighborhood size to 7.
	Q11: Is more of the noise removed by each smoothing filter? Is there noticeable loss of detail in the statue features compared to the original image?
	Q12: Do the edges in the median filtered image seem to move? Why does this happen?
	Q13: Why does a median filter with a neighborhood size of 7 do a better job of reducing noise particles of size 3 than a median filter with a neighborhood of size 3?
	Q14: Which image gives you the best idea of what the original image was - the noisy image, the median filtered image, or the averages image?
Step 8:	Set the neighborhood size back to 3.
Step 9:	Stop the worksheet. Close the worksheet.

Overview Questions

A: Why do smoothing filters have such different effects on different images? How does this depend on what is in the image?

B: When images have noise, how can we decide how much smoothing we want to do to reduce the appearance of the noise?

C: When is there the most difference between how the median filter changes the noise and how the averaging filter changes the noise?

Summary

Median and averaging filters can reduce the appearance of imperfections in an image, but they also change the appearance of edges and structures in the image. The amount of smoothing is controlled by the neighborhood size. The choice of the neighborhood size be based on a trade-off between how much noise reduction is needed and how much change in the underlying image content is acceptable.

4.8   Design of an Object Counter

Laboratory Objectives

In this laboratory you will design and test an object counter. We will start our design process with a definition of what we want the object counter to do and then build and test a worksheet based on our design. We will learn the value of dividing the big design objective into smaller functional units so we can work on several simple design problems instead of one really complicated problem. We will also learn the value of testing the project after each functional unit is added to the design so that it is easier to find and fix problems. A worksheet with suggested parts is supplied, but additional parts from the libraries may be used in the final design. We will also explore the behavior and uses of a object counter.

Textbook Reading

*   This lab appears on page **241** of the *Engineering Our Digital Future* textbook.
*   Prerequisite textbook reading before completing this lab: pp. **239-241**.

Engineering Designs and Resources

Worksheets used in this lab:

*   **L04-08-01 Design of an Object Counter Parts.Lst:** Students create a worksheet that will count objects in an image.

4.8.1   Design of an Object Counter Parts

Worksheet Description

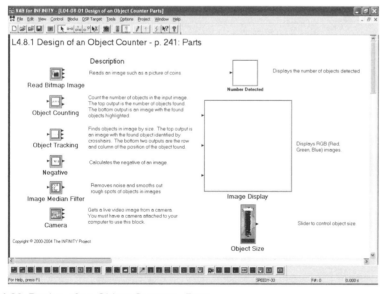

Figure 4.22 Design of an Object Counter - Parts

Laborator Procedures

Part 1: Decide what the object counter should do.
Our first step is to decide what we would like an object counter to do and make a block diagram of functional units, which shows in a very simple way how we want our object counter to work. Then we can design each of the functional units using VAB blocks from the parts list or the libraries. Draw a block diagram of functional units for your object counter. You may be guided by the discussion of functional units below, but you may also create a design using different functional units.

- If we want to count objects, we will need an input image with the objects to be counted. So our first functional unit will provide the image and display it. Since separating objects from the background may be a complex task, we will start by using a test image using a bitmap read block instead of a camera block. Later, when the object counter is working well, we can switch the input to a camera.
- The objects have to be separated from the background if we want to find them and count them. Our second functional unit will separate objects from the background using a threshold. We will use both a simple threshold block to explore the effect of changing thresholds and the more complex object counting block.
- We may want to adjust the image to make it easier for the Object Counting block to find objects in the image. Our third functional unit will be a set of blocks that will improve the image for use by the Object Counting block. This may not result in an image that is nicer for us to look at, but it will make the object counter more reliable.
- If we want the object counter to work in a variety of applications, we will probably want to easily control the object counting block. Our fourth functional unit will be a parameter control unit for the object counter. We will add sliders to control all the parameters.

Part 2: Design and test the functional units.
It is important to start with a simple design and test it, and then add more complicated features. If we connect a lot of blocks to make a complex design and then start the worksheet, it can be difficult to find problems and correct them. It is normal for the first design not to work the way we expected, and it is much easier to find a problem in a worksheet if we have just made a few changes to a worksheet that has been working correctly. We may often add image or numeric or graphical displays that are only used for testing. When the project is working, these extra displays may be removed or reused for other purposes.

Steps	Instructions
Step 1:	Open the worksheet *L04-08-01 Design of an Object Counter Parts.Lst*.
Step 2:	Open a new worksheet and save it with a new name. • Click **New** at the top of the pull down menu list for **File** at the top of your worksheet. A blank worksheet should appear on the screen. • Click **Save As** near the top of the pull down menu list for **File**. Navigate to the desktop. Your instructor may tell you what name should be used. In this description the new worksheet will be named *MyLab8*.

Important Note: You can quickly switch between your worksheet and the parts worksheet using **Window** at the top of your worksheet near **Help**. Click **Window** and look at the bottom entries in the pull down menu list. You should see the name you just used when you saved your new worksheet and you should see *L04-08-01*. You will also see the names of any other worksheets that are open. You can select any open worksheet by clicking its name in this list.

If you are using the functional units discussed above, then we can start with the first unit. If you are using different functional units, you should follow the overall procedure described here, but with specific details determined by your own design.

Unit 1: Read and display a test image

Steps	Instructions
Step 1:	We can get a test image input by putting Bitmap Read block in our worksheet and setting it up to read a suitable test image. • Go to the parts worksheet and copy the Bitmap Read block by clicking it to select it and then typing **Control C**. • Go to MyLab8 and paste the block by typing **Control V**.
Step 2:	To view the test image we will need an image display block. • Using the same method that we used in Step 1, copy the Image Display block from the parts worksheet and paste it into *MyLab8*.
Step 3:	Now we need to connect the Bitmap Read block to the image display so we can view the test image. • First select the **Bitmap Read** block by left clicking it, and drag it to the top left corner of the worksheet. • In the same way, place the Image Display block a little to the right of the bitmap read block. • Right click the label under the Bitmap Read block and change the name to "Read Image"
Step 4:	Select the **connect tool**. (It looks like a horizontal wire with connection points at each end.) • Connect the middle Bitmap Read block output (G) to all three inputs of the image display block.
Step 5:	Select the **setup tool**. (It is just to the left of the connect tool and looks like a diagonal arrow.) Right click the **Bitmap Read** block and make sure that the file name is *COIN2BWX288x300.bmp*. If it is a different name, then click **Browse** and: • Use the browser to navigate to Program Files\Hyperception\VAB-INF\Images\BWimages and select *COIN2BWX288x300.bmp* in this directory. • Click **Open** in the browser's Open window. • Then click **OK** in the Bitmap Read Parameters Window.
Step 6:	**Testing Unit 1:** Save your worksheet. Start your worksheet.

Steps	Instructions (Continued)
Step 7:	We should see a test image showing eight coins on a light background. If this is not working, then this must be fixed before we can go to the next unit.
Step 8:	Stop your worksheet.

Unit 2: Separate objects from the background.

We have already explored thresholding in Lab 4.2. Since the background is much lighter than the coins, we can try a threshold operation first. From the description of the Object Counting block operation, it sounds like it uses a threshold block. For this unit we will use both a Threshold block and the Object Counting block block.

Steps	Instructions
Step 1:	On the parts worksheet, right click the **Object Counting** block and select **Help**. Read the description of the block's function at the bottom of the help page.
	Q1: Does this block sound useful for our needs?
	Q2: What are the block's inputs?
	Q3: What are the block's outputs?
	Q4: What are the block's parameters?
Step 2:	Copy the Object Counting block and the Number Detected block from the parts worksheet to *MyLab8*. You can copy them at the same time by holding down the shift key when you left click them to select them. Then **Control C** will copy both of the selected blocks.

Steps	Instructions (Continued)
Step 3:	Add a Threshold block: From the Blocks drop-down menu on the top bar, click on **Select Blocks**. In the pop-up window that appears select **Image Processing Library** for the Library and for the Group List select **Arithmetic Functions**. • In the Function List on the right click **ImageThreshold** and then click the **Add to Worksheet** button. • Click the **Close** button to close the pop-up window. You should now see the new Threshold block on you worksheet.
Step 4:	Make two new copies of the image display block. • Left click the image display to select in and then type **Control C** to copy it. • Type **Control V** twice to paste two new displays into the worksheet.
Step 5:	Position the new blocks before connecting them. • Move the two new displays to the right side of the screen. Change the labels so that one is labeled "Detected Objects" and the other is labeled "Test". • Move the Object Counting Block and the Number Detected display to the middle of the worksheet just below the first Image Display. • Move the Threshold block to the lower left of the worksheet.
Step 6:	Select the **connect tool**. • Connect green image output from the Bitmap Read block to the input of the Objecting Counting Block and to the input of the Threshold block. • Connect the output of the Threshold block to all three inputs of the Test Display. • Connect the lower output of the Object Counting block to all three inputs of the Detected Objects display. • Connect the upper output of the Object Counting block to the input of the Number Detected display.
Step 7:	Select the **setup tool**. • Right click the **Threshold** block and make sure that the threshold value is 128, the True value is 255 and the False value is 0. • Right click the **Object Counting** block and record the parameter settings. The default settings should be minimum area = 7349, maximum area = 8674, and threshold = 70.
Step 8:	**Testing Unit 2:** Save your worksheet. Start your worksheet.

Steps	Instructions (Continued)
Step 9:	The Test image display should show the coins against a white background instead of the original light background for the image. However, with a threshold of 128, there are also many small white areas in the coins, so we have not completely separated the coins from the background. Try lowering the threshold to 100 and see what happens to the display.
	The Detected Objects display should not be working well. If we reread the Help information about the threshold for the Object Counting block, we see that pixels with values above the threshold will be part of the objects to be counted, not pixels below the value of the threshold. The Object Counting block is looking for light objects on a dark background and our test image has dark objects on a light background.
Step 10:	Stop the worksheet and plan a method to fix the problem we have just discovered.

Unit 3: Add image preprocessing to make the Object Counter work better.

Steps	Instructions
Step 1:	To make the object counter work better, we will need to invert the image so the coins will be lighter than the background. It could also help to have a smoother image so the small bright highlights of the coins are not confused with the background.
Step 2:	Copy the Negative block and the Image Median Filter block from the parts worksheet to *MyLab8*.
Step 3:	Position the new blocks before connecting them. • Move the Median Filter block to a position just to the left of the Object Counting block and just below the wire connected to the input of the Object Counting block. • Move the Negative block to a position just to the left of the Median Filter block.
Step 4:	Select the **delete tool**. (It looks like a pair of scissors.)Left click the wire connected to the input of the Object Counting block to delete the wire.
Step 5:	Select the **connect tool**. • Connect green image output from the Bitmap Read block to the input of the Negative Block. • Connect the output of the Negative block to the input of the Image Median filter block. • Connect the output of the Median Filter block to the input of the Object Counting block.
Step 6:	Testing Unit 3: Select the **setup tool.** Save your worksheet. Start your worksheet.

Steps	Instructions (Continued)
	Q1: How many coins are detected? Which ones are they?
	Q2: Does it look like the coins that are not detected are missed because the threshold is set two low or because the minimum and maximum area are not set correctly? (The image has a width of 288 pixels and a height of 300 pixels.)
Step 7:	Right click the **Object Counting** block and change the minimum area from 7349 to 2000. Then click **OK.**
	Q3: Now how many coins are detected?
Step 8:	Right click the **Object Counting** block and change the maximum area from 8674 to 5000. Then click **OK**.
	Q4: How many coins are detected? Which ones are they?
Step 9:	Right click the **Object Counting** block and change the threshold from 70 to 150. Then click **OK**.
	Q5: What happens to the coins at this higher threshold value?

Steps	Instructions (Continued)
Step 10:	Stop the worksheet.
	It will be much easier to explore the effect of the parameter settings if we add sliders so that we do not have to go through the procedures of Steps 7-9 to change values.

Unit 4: Add sliders to control the Object Counting block

Steps	Instructions
Step 1:	Now that we are more confident about the operation of the Object Counting Block, we can delete the Test display and the Threshold block to make room for the sliders.
	Select the **delete tool**. (It looks like a pair of scissors.)
	• Left click the **Test** display.
	• Left click the **Threshold** block.
	Select the **setup tool**.
Step 2:	Left click the upper left corner of the Object Size slider block from the parts worksheet to select it and then copy it. Paste it three times into *MyLab8*.
Step 3:	Drag the three sliders to the bottom center area of the worksheet and relabel them "Min Area", Max Area" and "Threshold".
Step 4:	Select the **parameter connect tool**. (It looks like a reflected "L" with connection points at both ends.)
	• Connect the Min Area slider to the Objecting Counting block and select the minimum area parameter to be controlled by the slider.
	• Connect the Max Area slider to the Objecting Counting block and select the maximum area parameter to be controlled by the slider.
	• Connect the Threshold slider to the Objecting Counting block and select the threshold parameter to be controlled by the slider.
Step 5:	Select the **setup tool**.
	• Right click the **Min Area** slider and set Top Value = 10000.0, Bottom Value = 0, and Steps = 10001.
	• Right click the **Max Area** slider and set Top Value = 10000.0, Bottom Value = 0, and Steps = 10001.
	• Right click the **Threshold** slider and set Top Value = 255, Bottom Value = 0, and Steps = 256.
Step 6:	**Testing Unit 4:**
	Save your worksheet.
	Start your worksheet.
Step 7:	Adjust the threshold value slider until the 8 coins are distinctly separated from the background. Then adjust the area sliders until all eight coins are detected.

Steps	Instructions (Continued)
	Q1: What values did you find for the sliders? Are they the same as the values found by other students in the laboratory?
Step 8:	Experiment with other settings for the sliders. Some specific instructions for adjusting the sliders are found in the next laboratory.
Step 9:	Stop the worksheet. Close the worksheet

Part 3: Test and Improve your design

Now that we have a basic object counter working, we can test it and consider ways to improve it.

Steps	Instructions
Step 1:	Start your worksheet.
Step 2:	The default slider settings should be Min Area = 763, Max Area = 7845, and Threshold = 100. All 8 coins should be detected.
	Q1: Why do you think the Object Counting block has parameters for a minimum area and a maximum area? What might happen if the minimum area were zero and the background had a lot of variation in it?
Step 3:	Slowly increase the Min Area until one coin is missed and 7 are counted.
	Q2: What Min Area value caused the coin to be missed? Which coin is missed? Estimate the number of pixels in that coin (with the threshold set at 100) by finding minimum area values where it is detected and minimum area values where it is missed.
Step 4:	Increase the Min Area again until a second coin is missed and 6 are counted.

Steps	Instructions (Continued)
	Q3: Which coins are missed? Estimate the number of pixels in the second missed coin (with the threshold set at 100) by finding minimum area values where it is detected and minimum area values where it is missed.
Step 5:	Slowly reduce the Max Area until another coin is missed and 5 are counted.
	Q4: What Max Area value caused the third coin to be missed? Which coin is it? Estimate the number of pixels in that coin (with the threshold set at 100) by finding maximum area values where it is detected and maximum area values where it is missed.
Step 6:	Set the minimum are to 3000, the maximum are to 7500, and the threshold to 100. Verify that all eight coins are counted.
Step 7:	Change the threshold to 160 and observe the results.
	Q5: How many coins are counted? Which ones are missed? Why are they missed?
Step 8:	Slowly reduce the minimum area to try to count the missed coins.
	Q6: Why doesn't this work very well? How many objects are counted when the minimum area is 0? Why?
Step 9:	Set the minimum area to 3000 and slowly reduce the threshold to 50. Observe the results.

Steps	Instructions (Continued)
	Q7: How many coins are counted? Which ones are missed? Point out two things that happened when the threshold was reduced that caused the coins to be missed.
Step 10:	Test your answer to Question 7 by raising the Max Area until some of the missed coins are counted.
	Q8: Which missed coins can be counted by raising the Max area value?
	Q9: Which coins are still missed even when Max Area is set to 10000? Why are they still missed even though they are smaller than the largest coins? For a threshold of 50, what could you change so that these coins would not be missed?
Step 11:	Stop the worksheet. Close the worksheet.

Improvements to the design:

- Consider replacing the negative block with the mapping function block from Lab 4.2. Could this be used to provide better input for the object counter?
- Think of ways to improve the way your object counter works. Think of new features to add. Try them and see what happens. Remember that most things don't work quite right the first time you try them, so be prepared to make adjustments and try again.
- Connect a camera in place of the image read but keep in mind that:
  - The Object Counting block was designed to operate correctly for specific types of input images, and with the camera you may create images that cause the block to give unusual outputs. (Remember the output we saw in the middle of Lab 4.1.1 when we connected the image directly to the Object Counting block without using the Negative block.

- The Object Counting block counts light objects on a dark background. We added the Negative block because the objects in our test image were dark objects on a light background. If you are trying to count light objects on a dark background, you should remove the Negative block. You could also replace the Negative block with a mapping block from Lab 4.2 so that you can flexibly change the input mapping of the camera image to best match the Coin Counting block.

If you use the mapping block, it will be helpful at add an image display to check that the output of this block is really what you want.

4.9   Design of a Motion Detector

Lab Objectives

In this laboratory you will design and test a motion detector. We will start our design process with a definition of what we want the motion detector to do and then build and test a worksheet based on our design. We will learn the value of dividing the big design objective into smaller functional units so we can work on several simple design problems instead of one really complicated problem. We will also learn the value of testing the project after each functional unit is added to the design so that it is easier to find and fix problems. A worksheet with suggested parts is supplied, but additional parts from the libraries may be used in the final design. We will also explore the behavior and uses of a motion detector.

Textbook Reading

*   This lab appears on page **242** of the *Engineering Our Digital Future* textbook.
*   Prerequisite textbook reading before completing this lab: pp. **240-241.**

Engineering Designs and Resources

Worksheet used in this lab:

*   **L04-09-01 Design of a Motion Detector Parts.Lst:** Contains the blocks needed to build the motion detector.

Worksheet Description

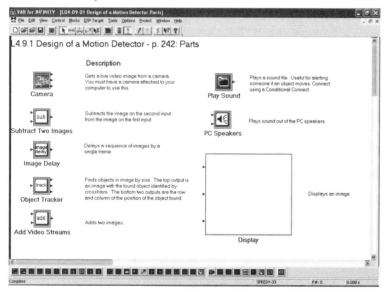

Figure 4.23 Design of a Motion Detector - Parts

Open the worksheet *L04-09-01 Design of a Motion Detector Parts.Lst.* You should see a screen that looks like the Figure 4.23 above with a selection of suggested parts for a motion detector.

Read the brief description of all the blocks on the worksheet.

Laboratory Procedure

Part 1: Decide what the motion detector should do.

Our first step is to decide what we would like a motion detector to do and make a block diagram of functional units which shows in a very simple way how we want our motion detector to work. Then we can design each of the functional units using VAB blocks from the parts list or the libraries. Draw a block diagram of functional units for your object detector. You may be guided by the discussion of functional units below, but you may also create a design using different functional units.

- If we want to detect motion, we will need two images - one current and one older. So our first functional unit will provide these two images. Its input will be the current image from the camera and its output will be two images - the current image and an older image from the camera.
- If any object moved after the older image was stored, the old image and the current image will be different. Our second functional unit will detect motion by comparing the two images and finding where they are different. Its input will be two images and its output will be something that tells us if the images are different. If we want a smarter detector, it might also tell us how different the images are or where the biggest differences are.
- A motion detector is not very useful if it detects motion, but does not communicate that to us. We will need a third functional unit that takes the output of the second block and communicates to us visually or through audio that something has moved.

Part 2: Create, Design, and test the functional units.

It is important to start with a simple design and test it, and then add more complicated features. If we connect a lot of blocks to make a complex design and then start the worksheet, it can be difficult to find problems and correct them. It is normal for the first design not to work the way we expected, and it is much easier to find a problem in a worksheet if we have just made a few changes to a worksheet that has been working correctly. We may often add image or numeric or graphical displays that are only used for testing. When the project is working, these extra displays may be removed or reused for other purposes.

Steps	Instructions
Step 1:	Open a new worksheet and save it with a new name. • Click **New** at the top of the pull down menu list for **File** at the top of your worksheet. A blank worksheet should appear on the screen. • Click **Save as** near the top of the pull down menu list for **File**. Navigate to the desktop. Your instructor may tell you what name should be used. In this description the new worksheet will be named *MyLab9*.

You can quickly switch between your worksheet and the parts worksheet using Window at the top of your worksheet near Help. Click **Window** and look at the bottom entries in the pull down menu list. You should see the name you just used when you saved your new worksheet and you should see L04-09-01. You will also see the names of any other worksheets that are open. You can select any open worksheet by clicking its name in this list.

If you are using the functional units discussed above, then we can start with the first unit. If you are using different functional units, you should follow the overall procedure described here, but with specific details determined by your own design.

Unit 1: Create delayed image

Steps	Instructions
Step 1:	We can get an image input by putting a camera in our worksheet. • Go to the parts worksheet and copy the camera by clicking it to select it and then typing **Control C**. • Go to *MyLab9* and paste the camera by typing **Control V**.
Step 2:	To make a delayed image we can use the image delay from the parts worksheet. It will give us a delayed image for a single component, so we probably should have three image delays, one for each color component. However, in previous laboratories we have found that for most color images the green component provides image information very similar to a grayscale image, so in our initial design we will just use the green component. • Copy the image delay block from the parts worksheet and paste it into *MyLab9*.
Step 3:	**Testing Unit 1:** We can just connect the green component output from the camera to the input of the image delay and we will have built our first functional unit. But how will we know it is working? • We can copy the image display from the parts worksheet and paste it into our current worksheet twice. • Select the image display on the parts worksheet and copy it. • Paste two displays onto *MyLab9*.
Step 4:	Select the **connect tool**. • Connect the three R, G, and B outputs from the camera to the corresponding three inputs of the first image display. • Connect the middle camera output (G) to the image delay block input. • Connect the image delay output to all three inputs of the second image display.
Step 5:	Select the **setup tool**. Save your worksheet. Start your worksheet.
Step 6:	We should see the live camera image in a color display and a delayed grayscale image of the green component in a grayscale display. Place your hand in front of the camera so you can see your fingers moving. Do you see a delayed movement in the grayscale display? If this is not working, then this must be fixed before we can go to the next step.

Unit 2: Detect change

We have already explored a change detector in the extension to Lab 4.3. In the basic laboratory we could see the difference of two images taken at different times because one was subtracted from the other. When we looked at the difference image, we could see the change. When we used a camera in the extension, the difference image was black unless something moved in front of the camera. But that will only be the first part of our change detector. We want an automatic change detector, not just a device that makes it easier for us to notice when a change happens.

Steps	Instructions
Step 1:	On the parts worksheet right click the **object tracker** block and select **Help**. Read the description of the object tracker's function at the bottom of the help page. Does this block sound useful for our needs?
Step 2:	Copy the object tracker block, the subtract block, and the add block from the parts worksheet to *MyLab9*. You can copy them all at the same time by holding down the shift key when you click them. **Control C** will copy all the selected blocks.
Step 3:	Select the **connect tool**. • Connect the delayed green image output to the bottom input of the subtract block. • Connect the green camera output to the top input of the subtract block. Now the output of the subtract will be the difference between the current image and an older image. • Connect the subtract block output to the input of the object tracker block so the object tracker can analyze this difference.
Step 4:	**Testing Unit 2:** Once again, we have created a functional block that we think will do what we want, but we need to test it. We can do that by reusing the display of the delayed image because we do not need that display for our automatic motion detector.
Step 5:	Select the **delete tool**. (It looks like a pair of scissors.) • Delete the three inputs to the second image display by clicking them.
Step 6:	Select the **setup tool**. Drag the now disconnected second image display to the right of the worksheet.
Step 7:	Select the **connect tool**. • Connect the red and blue camera outputs to the corresponding inputs of the second display. • Connect the green camera output to the top input of the add block • Connect the top object tracker output to the lower input of the add block. Now the add block output will be the sum of the green color component and the object tracker information about the location of the motion. • Connect the adder output to the green input of the second display.
Step 8:	Select the **setup tool**. Save your worksheet. Start your worksheet.
Step 9:	Put your hand about three inches from the camera and move it slowly. What do you see?

Steps	Instructions (Continued)
Step 10:	Test your motion detector with a variety of inputs. Aim the camera at things that are close to it and at things that are distant from it.

Unit 3: Make an audio alert

Steps	Instructions
Step 1:	On the parts worksheet, right click the block labeled "Play Sound" and select **Help**. At the end of the help sheet, read the description of the File Read block.
Step 2:	Copy the "Play Sound" file read block and the speaker block to *MyLab9*.
Step 3:	Connect the Play Sound output to both speaker inputs.
Step 4:	Use the **conditional connect tool** (it looks like the connect tool with a question mark) to connect the second object tracker output to the Play Sound Block.
Step 5:	Select the **setup tool**. Save your worksheet. Start your worksheet.
Step 6:	Now what happens when something moves in front of the camera?

Part 3: Test and Improve your design
Now that we have a basic motion detector working, we can test it and think of ways to improve it.

Laboratory Procedures

Steps	Instructions
Step 1:	Start your worksheet.
	Q1: What happens when something moves in front of the camera?
	Q2: Can you move your hand across the camera without setting off the motion detector? How can you do this? How can you improve this motion detector design to make what you did harder?

Steps	Instructions (Continued)
	Q3: Hold the camera in your hand and point it at a bookshelf full of books or window blinds or some other scene with a lot of different objects. Can you hold the camera still enough so that no motion is detected?
	Q4: How still would you have to hold the camera so that no motion would be detected? Would it depend on the distance of the objects in the camera field of view?
	Q5: Could this motion detector be used as a vibration detector?
Step 2:	Explore the parameter settings for object tracker and tune them for a particular motion detection application.
	Q6: We are only detecting motion of the green component. Is this likely to ever cause a problem? (Remember that white has a strong green component and black has a zero green component.)
	Q7: What would you have to do to make the motion detector shout "STOP" when it detected motion?

Possible improvements to the design include:

- Explore the parameter settings for the object tracker block and tune them for a particular motion detection application.
- Think of ways to improve the way your object tracker works. Think of new features to add. Try them and see what happens. Remember that most things don't work quite right the first time you try them, so be prepared to make adjustments and try again.

4.10 Design of a Blue Screen

Lab Objectives

In this laboratory you will experiment with chromakey methods which are used in television and movies to create special effects. You can put yourself in the middle of any stored image in the same way that a weatherperson on television appears to be in front of a huge weathermap. You can add only your hands but not your arms and the rest of your body to an image to make it look like the hands are completely independent. The secret to these effects is the selective combination of images from two sources. The image that includes you will have a range of colors that are made transparent when the images are combined. You will use techniques from many of the earlier laboratories in this chapter to combine images and understand the structure of the color cube.

Textbook Reading

*   This lab appears on page **244** of the *Engineering Our Digital Future* textbook.
*   Prerequisite textbook reading before completing this lab: pp. **242-244**.

Engineering Designs and Resources

Worksheet used in this lab:

*   **L04-10-01 Design of a Blue-Screen Image System.Lst:** Performs "chromakey" color replacement processing on a live video stream replacing a defined sub-cube with a fixed image of a weather system.

Worksheet Description

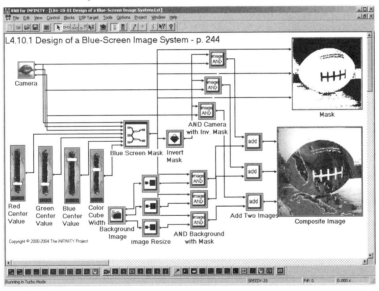

Figure 4.24 Design of a Blue-Screen Image System

Open the worksheet *L04-10-01 Design of a Blue-Screen Image System.Lst.* You should see a screen that looks like Figure 4.24 above except that the displays will be blank.

The foreground image comes from the camera in the upper left part of the worksheet. Some colors in this image will be chosen to become transparent. Pixels with other colors will not be transparent and will appear in the composite image in the lower right. These pixels appear in black in the mask image in the upper right. Pixels that are chosen to be transparent appear as white in the mask in the upper right. In the composite image, transparent pixels are replaced by pixels from the same location in the background image

The colors that are chosen to be transparent are controlled by four sliders in the lower left part of the worksheet. How these sliders choose colors will be described in some detail later. The Blue Screen Mask hierarchy block in the center of the worksheet uses the foreground image and the selection of transparent colors to make the mask image.

- The mask, which is white for transparent pixels, is logically ANDed with the background image so that the background image will replace transparent pixels.
- The negative of the mask is logically ANDed with the foreground image so that pixels which are not transparent will come from the foreground image.

These two images are then added to form the composite image in the lower right.

In a typical application the foreground image would be an image of you in front of a blue screen. Blue pixels would be chosen to be transparent, so all parts of the image except you would come from the background image. Any color may be chosen for the background, but bright blue is often used because it is least likely to appear in skin tones. Bright green may also be chosen for the background color.

In the example in Figure 4.24 above, blue is a poor choice for transparent pixels because the foreground image is a blue football on a wooden table with a light background. The blue football becomes transparent revealing the background weather map in the composite image. The white laces of the toy football and the light background are not transparent, so they appear in the composite image.

For this laboratory it is important to decide what color will be the background color and then make sure that none of the important foreground image content, usually you and your clothing, have this color. The background should be as uniform in color as possible and have even lighting to be most effective.

How can we select the transparent pixels?

If we want to have bright blue pixels be background pixels, we can not just limit our attention to the blue color component of the image. White pixels have as strong a blue component and bright blue pixels. One way to find only blue pixels is to define a subcube of the color cube as shown in the figure below. We want pixels near the blue corner of the cube to be transparent, so we could say that all pixels with blue values between 245 and 255 AND red values between 0 and 9 AND green values between 0 and 9 will be transparent. Any pixel with a color outside that range for any of the color components would be a foreground pixel and would appear in the composite. Note that white and cyan and magenta all have a blue value of 255, but would not be found in this sub-cube,

The choice of the size and location of the sub-cube will depend on the variation in the color and the lighting of the background. For example, light blue might have R=128, G=128, and B=255. This would be outside the small sub-cube we used for our example.

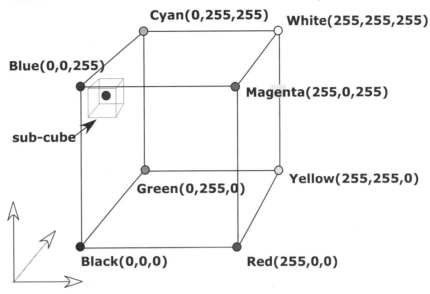

Figure 4.25 Color Cube and Sub-cube as chosen by 8-bit color plane values

The four sliders in the worksheet will define this sub-cube. Three sliders control the red, green, and blue values of the center of the cube. The fourth slider controls the size of the cube. Based on the slider values, threshold blocks can be used to select pixels in the sub-cube and then logical AND operations can combine the conditions to make the mask.

The actual implementation uses the center value plus the cube width for the high threshold and the center value minus the cube width for the low threshold. Knowing this makes some parts of the laboratory easier to understand. This means that the slider value for color cube width is really only half the width. The other important note is that it does not matter if the threshold values are below 0 or above 255.

Steps	Instructions
Step 1:	Set up your blue screen studio. Find a uniform blue or green background so that when students stand in front of it, the background can be made transparent.
Step 2:	**Open** the worksheet *L04-10-01 Design of a Blue Screen Image System.Lst.* Start the worksheet. One student, who is not wearing the same color clothing as the background, should stand in front of the background while another student adjusts the slider controls on the worksheet.

Steps	Instructions (Continued)
	Q1: What do you observe? With the initial slider settings of the sub-cube center values (red = 46, green = 54, and blue = 220) and cube width = 158 what parts of the background are transparent ? Can you see your lab partner in front of the background image of a hurricane weather map?
Step 3:	Set the cube width to 255. You should see the complete weather map in the composite image and the mask should be completely white.
Step 4:	Set the cube width to 0. Now you should see the complete foreground image in the composite image because the size of the sub-cube is 0. The mask image should be completely black.
Step 5:	Using the cursor, click several pixels in the parts of the foreground image (your lab partner in front of the blue screen) that should become transparent and record their RGB values.
	Q2: Compute the following: -the average red value of all the pixels you recorded -the average blue value -the average green value -the largest difference between any color component value and its average value
Step 6:	Set the center value sliders to the average values you computed in Question 2. Set the cube width to the largest difference you computed.
	Q3: How does the composite image look now? Are all the pixels in the blue screen replaced by the weather map in the composite image? Are any parts of your lab partner missing from the composite image?

Steps	Instructions (Continued)
	Q4: Make a sketch like Figure 4.25 which shows your sub-cube from Step 6 on the color cube. It is ok if your sub-cube extends beyond the boundaries of the color cube.
Step 7:	Make fine adjustments to improve your composite image. You may need to make some compromise between making blue screen pixels transparent and retaining pixels belonging to the student in front of the blue screen.
	Q5: Explain how the mask image is used to make the composite image.
Step 8:	While using the camera based blue screen demo have the person in the video point to a feature on the image (Point to Florida if you are using a weather map). Where are they looking as they are pointing? Does this remind you of the weather forecaster?
Step 9:	Experiment with other effects. For example, have a student wearing a blue shirt stand in front of the blue screen. Try to adjust the sliders so you only see the hands moving by themselves in front of the weather map.
Step 10:	Stop the worksheet. Close the worksheet.

Quantitive Investigations:

It is a little difficult to analyze the blue screen system with the camera input because the variations in light and motion of the subjects makes it hard to reproduce conditions for testing. In this section we will replace the camera with a Bitmap Read block and use a test image in place of the camera input.

Steps	Instructions
Step 1:	Open the worksheet *L04-10-01 Design of a Blue Screen Image System.Lst*.
Step 2:	Note that each camera output is connected to two inputs - the bottom input of an image AND block and an input to the hierarchy Blue Screen Mask block.
	With the setup tool selected, left click each of these wires one at a time to highlight them, and note how they are connected.
Step 3:	Delete the camera. Select **delete tool**. Left click the camera to delete it and the wires connected to it.
Step 4:	Add a Bitmap Read block.
	Select the **setup tool**.
	From the Blocks drop-down menu, click **Select Blocks**.
	Select the **Image Processing** library. For the group list, select File I/O functions. Under the Function List click **Bitmap read**.
	Click **Add to Worksheet** and then **Close**.
Step 5:	Drag the Bitmap Read block to the position where the camera was.
	Right click it, and with browser navigate to Program Files\Hyperception\VAB-INF\Images\colorImages and select *stickPerson352x288.bmp* in this directory.
	Click **Open** and then **OK**.
Step 6:	Right click the Bitmap Read label and change it to "Test Image"

Steps	Instructions (Continued)
Step 7:	Connect the Bitmap Read block as the camera was connected. Select connect tool. • Connect the top (R) output of the Bitmap Read block to the open lower input of the top AND block and to the top open input of the Blue Screen Mask block. • Connect the middle (G) output of the Bitmap Read block to the open lower input of the next AND block and to the next open input of the Blue Screen Mask block. • Connect the lower (B) output of the Bitmap Read block to the open lower input of the next AND block and to the next open input of the Blue Screen Mask block. Select the setup tool. Your worksheet should now look like Figure 4.26. 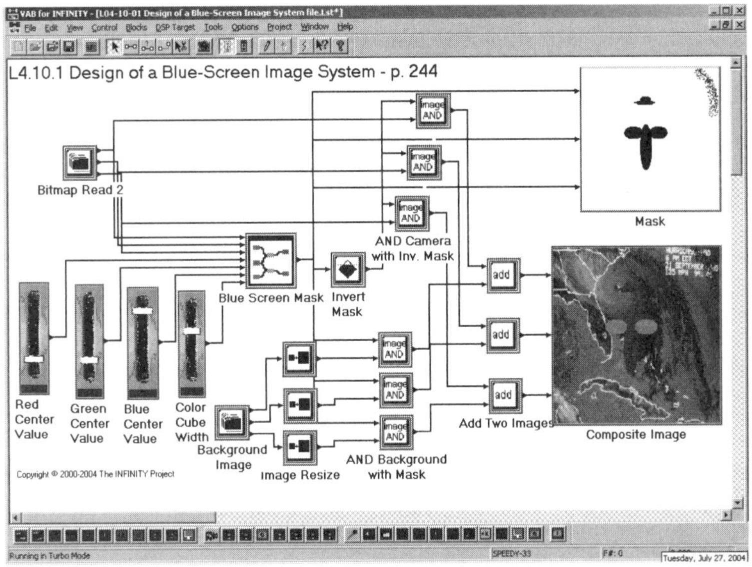 **Figure 4.26** Blue screen worksheet with test image
Step 8:	Save the worksheet as *MyTestLab10* on the desktop.
Step 9:	Start the worksheet.
Step 10:	With initial center values of 46, 54, 220, and an initial width of 158, the composite in Figure 4.26 shows just the hat, arms, and body of the stick person superimposed on the hurricane weather map.
Step 11:	Change the color cube width to 0. Now the whole mask is black and only the test image is seen in the composite image.

Steps	Instructions (Continued)
	Q1: Record the RGB values of the following parts of the stickperson image. Use the cursor to click pixels in the image and see the RGB values.

Stickperson	red	green	blue
Legs			
Body			
Arms			
Head			
Hat			
Blue background			

	Q2: Using the slider values in step 10, compute the ranges of the sub-cube components in the background. The low value for each component is the center value minus the width. The high value of a component is the center value plus the width.

Component	low value	high value
Red		
Green		
Blue		

Steps	Instructions (Continued)
	Q3: From your answers to Questions 1 and 2, explain why the blue background and the legs and head of the stickperson did not appear in the composite image in Step 10.
Step 12:	Decrease the width of the sub cube slowly from 255 to 0 with the slider.
	Q4: Note the value for the width at which various components of the test image lose transparency and become visible in the composite image. In each case, state whether it was the red, green, or blue component that fell outside the sub-cube of transparent colors.
Step 13:	Find settings of the sliders that will show the whole stickperson in the composite but will show as little of the blue background as possible. **Figure 4.27** Complete stickPerson.
	Q5: Record your settings and compare them with other students' results.
Step 14:	Experiment with other settings that will make red areas transparent but keep blue. Where should the center of the sub-cube be? Test it with the test image.
Step 15:	Continue to explore with a second test image, *blueOvals352x288.bmp,* which can be found in the directory Program Files\Hyperception\VABINF\Images\colorImages.

Steps	Instructions (Continued)
Step 16:	Stop the worksheet. Close the Worksheet.

Optional Further Explorations: How does the Blue Screen Mask Block work?

Steps	Instructions
Step 1:	Open the worksheet *L04-10-01 Design of a Blue-Screen Image System.Lst*. Start the worksheet. Set the center values to red = 40, green = 60 and blue = 200. Set the width to 100. Stop the worksheet.
Step 2:	Double left click the **Blue Screen Mask** block to view its component parts. Many blocks will appear to be on top of one another, but you can drag them apart to make the worksheet look like Figure 4.28 below. 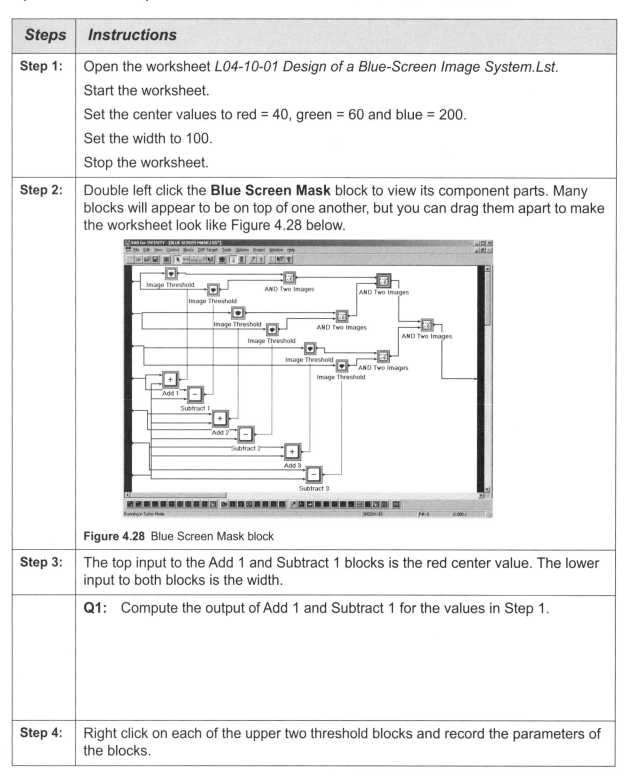 **Figure 4.28** Blue Screen Mask block
Step 3:	The top input to the Add 1 and Subtract 1 blocks is the red center value. The lower input to both blocks is the width.
	Q1: Compute the output of Add 1 and Subtract 1 for the values in Step 1.
Step 4:	Right click on each of the upper two threshold blocks and record the parameters of the blocks.

Steps	Instructions (Continued)
	Q2: How are the threshold levels related to the values you computed in Question 1? Note that one threshold block outputs a 0 when the input value is greater than its threshold and the other outputs a 255 when its input value is greater than its threshold.
	Q3: The input to these two threshold blocks is the red pixel value. What red pixel values will cause the output of both of the threshold blocks to be 255? What part of the color cube would this be?
Step 5:	Repeat Steps 3 and 4 for Add 3 and Subtract 3 blocks and the lowest two threshold blocks. Find the values for the blue color component that will make the output of both of these threshold blocks 255.
	Q4: What part of the color cube would cause both these outputs to be 255?
	Q5: What part of the color cube will cause the two threshold blocks from Question 3 and the two threshold blocks from Question 4 to all have an output of 255?
Step 6:	Explore the thresholds for the green input in the same way.
Step 7:	Click the vertical blue arrow icon to move back up to the original worksheet.
Step 8:	Stop the worksheet. Close the Worksheet.

Overview Questions

A: What happens if the color you choose to make transparent is prevalent in your clothing or face?

B: What would happen if you were wearing clothing which matched the color of the back-ground?

C: How can you determine which pixels are in the cube and which are not?

Summary

Chromakey replaces a region of the color cube in a live video stream with a fixed image, like a weather map.

Digitizing the World

Chapter 5 focuses on the general ideas associated with the digitization of a wide range of information from text in books and magazines, to speech and music, to images and video. Students learn the details of how all these types of information are captured and stored in digital form on a computer. They also learn about the various practical trade-offs when real world or "analog" information is converted to numbers and stored with finite precision.

Infinity Labs

5.1 Aliased Sinusoids, Speech, and Music

5.2 Quantization and Clipping

Introduction

The laboratories in Chapter 5 will focus on the issues surrounding the sampling and quantization of sound signals as they are converted into digital form. Sampling refers to how often a number is recorded for the input signal, whereas as quantization refers to the accuracy of the number that is recorded. The effects of sampling above and below the Nyquist Sampling rate are examined as well as the effects of introducing noise through less accurate quantization requiring fewer bits. There is a trade-off between the quality of the recorded signal and the number of bits/second it requires as defined by the sampling rate (number of samples per second) and the quantization (bits per sample).

The basic input blocks of Chapter 2, the microphone and Cosine generators, will be used in this chapter again. The output blocks are similarly the signal display and speakers. The main focus of this Chapter will be altering the sampling rate and quantization levels of the signals through the use of sliders.

5.1 Aliased Sinusiods, Speech, and Music

Lab Objectives

In the previous chapters, we have dealt with many different signals (most are based on cosines) but have not looked at how the computer can process them. In this chapter we see how the signals get recorded in the computer before it can process them. There are two steps to this recording process. The first is to read the value of the waveform ever so often. This is called sampling. In this lab we will have the DSP sample at either 8,000 samples per second or 44,100 samples per second to determine the effects of sampling frequency.

Textbook Reading

- This lab appears on page **263** of the *Engineering Our Digital Future* textbook.
- Prerequisite textbook reading before completing this lab: pp. **252-263.**

Engineering Designs and Resources

Worksheets used in this lab:

- **L05-05-01 Aliased Sinusoids Speech and Music Sine Wave.Lst:** Students learn about the effects of temporal aliasing of sinusoids.
- **L05-01-02 Aliased Sinusoids Speech and Music Square Wave.Lst:** Students learn about the effects of temporal aliasing of square waves.
- **L05-01-03 Aliased Sinusoids Speech and Music Microphone.Lst:** Students learn about the effects of temporal aliasing of other sound signals. Aliased Sinusiods, Speech, and Music - Sine Wave

5.1.1 Aliased Sinusoids, Speech, and Music - Sine Wave

Worksheet Description

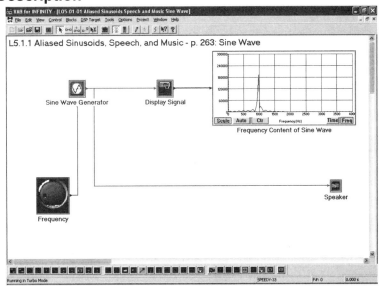

Figure 5.1 Aliased Sinusoids, Speech, and Music - Sine Wave

Open the worksheet *L05-01-01 Aliased Sinusoids Speech and Music Sine Wave.Lst.*You should see something similar to Figure 5.1 above.

This worksheet will allow you to listen to sinusoids of various frequencies using different sampling rates.

Laboratory Procedures

Steps	Instructions
Step 1:	Start the worksheet.
Step 2:	Then turn the **Frequency Knob** from 150 slowly up to 8,000.
	Q1: Describe what happens to the pitch of the signal as you increase the frequency of the input signal.
	Q2: At what frequency does the peak in the Frequency Content Waveform Display "turn around?" What happens to the sound as you continue to increase the frequency using the **Frequency Knob**?
Step 3:	Click **Stop**. Double click the **Frequency Knob** (this will open up a dialog box to allow you to change its parameters). Change Min Value to 0 and Steps to 801. This will allow you to change the frequency of the sinusoid in 10 Hz steps. Note: You may need to make the knob larger to get 10 Hz step.
Step 4:	Set the **Frequency Knob** to a value of 500 Hz.
Step 5:	Click the **Time** button on the Waveform Display to see the time varying signal.
Step 6:	Stop the worksheet.
Step 7:	Estimate the period of the waveform in the display. Note: For best results, measure the total time for 5 or 10 periods and then use the average value for one period. Use the cursor to read the time value.
	Q3: Using the measured period, what is the frequency of the waveform?

Steps	Instructions (Continued)
	Q4: How does your predicted frequency compare to the value of the **Frequency Knob** (500 Hz)?
Step 8:	Click the **Freq** button on the Waveform Display to see the frequency content of the signal.
	Q5: Where is the peak of the frequency content display?
	Q6: How does that compare with your computed value?
Step 9:	Set the **Frequency Knob** to a value of 7,500 Hz.
Step 10:	Click the **Time** button on the Waveform Display to see the time varying signal.
Step 11:	Stop the worksheet.
Step 12:	Estimate the period of the waveform in the display using the same method from Step 6.
	Q7: What frequency does your measured period predict the waveform is?
	Q8: How does your predicted frequency compare to the value of the **Frequency Knob** (7,500 Hz)?

Steps	Instructions (Continued)
Step 13:	Click the **Freq** button on the Waveform Display to see the frequency content of the signal.
	Q9: Where is the peak of the frequency content display?
	Q10: How does your answer compare with your computed value? How does this result compare to your previous result in Step 6?
Step 14:	Click **Go**, and turn the **Frequency Knob** to its maximum value of 8,000 Hz.
	Q11: Do you hear anything? Does this surprise you? Find other frequencies where you can not hear anything. Check low frequencies as well as high frequencies.
Step 15:	Now we will change the sampling rate from 8,000 to 44,100 Hz. • Click **Stop**. • Double Click the **Sine Wave Generator** block (this will open up a dialog box to allow you to change its parameters). • Click the **v** next to Sample Rate (another dialog box will open). • Double click the **8000** next to Sample Rate (a third dialog box will open). • Enter a value of 44,100 in place of 8,000 and click **OK**, **OK**, and **OK** to close the three nested dialog boxes. You have just changed the sampling frequency of the DSP board to 44,100 Hz. It is also necessary to change the frequency for the speaker. Right click on the speaker block. From the drop-down menu for sample rate, select **44,100**. Click **OK**.
Step 16:	Start your worksheet. Turn the **Frequency Knob** from 150 slowly up to 8,000.

Steps	Instructions (Continued)
	Q12: Describe what happens to the pitch of the signal. How does it compare to your answer to Question 1?
Step 17:	Set the **Frequency Knob** to a value of 7,500 Hz.
Step 18:	Click the **Time** button on the Waveform Display to see the time varying signal.
Step 19:	Stop the worksheet.
Step 20:	Estimate the period of the waveform in the display by counting several periods and taking the average.
	Q13: Using the measured period, compute the frequency of the waveform.
	Q14: How does your predicted frequency compare to the value of the **Frequency Knob** (7,500 Hz)?
Step 21:	Click the **Freq** button on the Waveform Display to see the frequency content of the signal.
	Q15: Where is the peak of the frequency content display?
	Q16: How does that compare with your predicted value?

Steps	Instructions (Continued)
Step 22:	Double click the **Frequency Knob**. (this will open up a dialog box to allow you to change its parameters). Change the maximum value from 8,000 to 20,000. Click **OK**. Now the **Frequency Knob** will increase the frequencies all the way up to 20,000 Hz or 20 KHz - near the human limit of hearing.
Step 23:	Start your worksheet. Turn the **Frequency Knob** from 150 slowly up to 20,000.
	Q17: Describe what happens to the pitch of the signal.
	Q18: Do you hear any aliasing of the signal? (Does the frequency ever "turn around" and start decreasing as you increase the frequency using the **Frequency Knob**?)
	Q19: At high frequencies the sound from the speakers has a reduced volume. Do you think this is due to limitations of the speakers or limitations in your hearing?
	Q20: At what frequency does the sound from the speakers begin to be reduced in volume? At what frequency do you have difficulty hearing it? Do you think this is due to the limitations of the speakers or your own ears?

5.1.2 Aliased Sinusoids, Speech, and Music - Square Wave

Worksheet Description

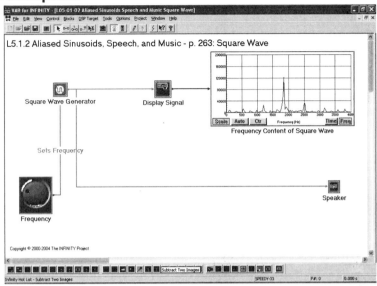

Figure 5.2 Aliased Sinusoids, Speech, and Music - Square Wave

Open the worksheet *L05-01-02 Aliased Sinusoids Speech and Music Square Wave.Lst.*You should see something similar to Figure 5.2 above.

This worksheet will allow you to listen to square waves of various frequencies sampled at different rates. In Lab 5.1.1 the sinusoids had only a single tone contained in them. In this lab the square waves will contain many tones (sinusoidal signals) so the effect is even more interesting.

Laboratory Procedures

Steps	Instructions
Step 1:	Double click the **Frequency Knob** (this will open up a dialog box to allow you to change its parameters).
	Change Min Value to 0 and Steps to 801. This will allow you to change the frequency of the sinusoid to 10 Hz steps.
	(Note: You may need to make the knob larger to get 10 Hz steps.)
Step 2:	Set the frequency to 150. For the sinusoid, we heard nothing here.
Step 3:	Why can we hear the square wave? (Hint: Look at the frequency display.)
Step 4:	Start your worksheet and turn the **Frequency Knob** to 400Hz.
	Stop the worksheet.
Step 5:	Click the **Time** button on the Frequency Content Waveform Display.

Steps	Instructions (Continued)
	Q1: Describe and sketch the waveform that you see. What is its period? What is its fundamental frequency?
Step 6:	Click the **Freq** button on the Frequency Content Waveform Display.
	Q2: Describe and sketch the waveform that you see. Identify the frequency values of any peaks in the plot using the cursor. How do they relate to the fundamental frequency?
Step 7:	Start the worksheet. Then turn the **Frequency Knob** from 0 slowly up to 500.
	Q3: Describe what happens to the waveform on the display. How does the spacing between the largest four peaks change as you increase the frequency?
Step 8:	Using the **Frequency Knob**, increase the frequency slowly from 500 Hz to 1,000 Hz.
	Q4: What happens to each peak as it "passes" 4 KHz on the display?
Step 9:	For each of the three frequencies 970 Hz, 1,000 Hz, and 1,030 Hz examine the frequency display and listen to the sound.

Steps	Instructions (Continued)
	Q5: Why is the sound at 1,000 Hz different from the sound at 970 Hz or 1,030 Hz? How is the frequency display different?
Step 10:	Set the frequency knob to a value of 810 Hz.
Step 11:	Click the **Time** button on the Waveform Display to see the time varying signal.
Step 12:	Stop the worksheet.
Step 13:	Estimate the period of the waveform in the display.
	Q6: What fundamental frequency does your measured period predict the waveform is?
	Q7: How does your predicted fundamental frequency compare to the value of the **Frequency Knob** (810 Hz)?
Step 14:	Click the **Freq** button on the Waveform Display to see the frequency content of the signal.
	Q8: Where is the peak of the fundamental frequency content displayed?
	Q9: How does your answer compare with your predicted value?

Steps	Instructions (Continued)
Step 15:	Change the frequency from 800 to 1,000. What happens to the harmonic? Why?
Step 16:	Set the frequency to 2,000. Why do all harmonics line up?
Step 17:	Now we will change the sampling rate from 8,000 to 44,100 Hz. • Click **Stop**. • Double click the **Sine Wave Generator** block (this will open up a dialog box to allow you to change its parameters), • Click the **v** next to Sample Rate (another dialog box will open). • Double click the **8000** next to Sample Rate (a third dialog box will open). • Enter a value of 44,100 in place of 8,000 and click **OK**, **OK**, and **OK** to close the three nested dialog boxes. You have just changed the sampling frequency of the DSP board to 44,100 Hz. It is also necessary to change the frequency for the speaker. Right click on the speaker block. From the drop-down menu for sample rate, select **44,100**. Click **OK**.
Step 18:	Start the worksheet. Then turn the **Frequency Knob** from 0 slowly up to 8,000.
	Q10: Describe what happens to the pitch of the signal as you increase the frequency. Do you observe the same effect that you heard before in Steps 8 and 9?
Step 19:	Set the **Frequency Knob** to a value of 7,500 Hz.
Step 20:	Click the **Time** button on the Waveform Display to see the time varying signal.
Step 21:	Stop the worksheet.
Step 22:	Estimate the period of the waveform in the display.
	Q11: What fundamental frequency does your measured period predict the waveform to have?

Steps	Instructions (Continued)
	Q12: How does your predicted fundamental frequency compare to the value of the **Frequency Knob** (7,500 Hz)?
Step 23:	Click the **Freq** button on the Waveform Display to see the frequency content of the signal.
	Q13: Where is the peak of the fundamental frequency content displayed? Use the cursor to measure its frequency.
	Q14: How does that compare with your predicted value?
Step 24:	Double click the **Frequency Knob**. (this will open up a dialog box to allow you to change its parameters). Change the maximum value from 8,000 to 20,000. Click **OK**. Now the **Frequency Knob** will increase the frequencies all the way up to 20,000 Hz or 20 KHz - near the human limit of hearing.
Step 25:	Click **Go**, then turn the **Frequency Knob** from 0 slowly up to 20,000.
	Q15: Describe what happens to the pitch of the signal.

Steps	Instructions (Continued)
	Q16: Do you hear any aliasing of the fundamental frequency? Do you hear any frequencies "descending" in pitch as you increase the value of the **Frequency Knob**?

5.1.3   Aliased Sinusoids, Speech, and Music - Microphone

Worksheet Description

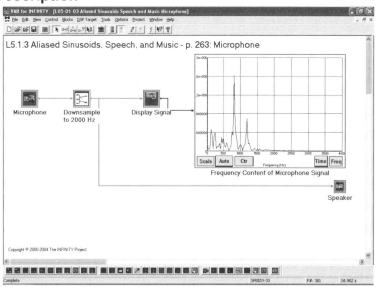

Figure 5.3 Aliased Sinusoids, Speech, and Music - Microphone

Open the worksheet *L05-01-03 Aliased Sinusoids Speech and Music Microphone.Lst.* You should see something similar to Figure 5.3.

This worksheet allows you to hear a version of your own or your lab partner's voice that has been downsampled to 2,000 Hz. This is the same type of downsampling that you used in Lab 3.2 with images.

Laboratory Procedures

Steps	Instructions
Step 1:	Start your worksheet.
Step 2:	Speak into the microphone.

Steps	Instructions (Continued)
	Q1: Describe how a voice sounds through this system.
	Q2: Can you understand what is being said?
Step 3:	Speak words with similar sounding letters: e.g., bed, pad, tad, ted; or words with the letter f and s.
	Q3: How well can you hear which letter is being said? Do you need a context to figure them out?
	Q4: Explain why certain letters are harder to understand than others. Can you relate this to the aliasing of sine waves and square waves?

Overview Questions

A: What is the definition of aliasing?

B: When does aliasing occur with respect to the sampling frequency?

C: What are the effects of aliasing on sinusoidal signals? On speech? On music? And on the combination of sinusoidal signals like square waves?

D: What would happen to the display and the sound of a sinusoidal signal sampled at its Nyquist frequency?

E: What would happen if the sampling rate were reduced to half the Nyquist frequency? (Defined on page 259 of the text.)

F: How does an aliased sinusoid sound? Can you notice the difference?

Summary

In this chapter we saw how the signals got recorded in the computer before it could process them. There were two steps to this recording process. The first was to read the value of the waveform ever so often. In this lab we had the DSP sampled at 8,000 samples per second and 44,100 samples per second to determine the effects of sampling frequency.

5.2 Quantization and Clipping

Lab Objectives

In the previous chapters, we have dealt with many different signals (most are based on cosines) but have not looked at how the computer can process them. In this lab, we see how the signals must be represented by numbers before the computer can process them. There are two steps to this recording process. The first is to read the value of the waveform every so often. This is called sampling. The second process is to convert the value that is read to an integer. This is called quantization. In this lab we will experiment with quantization.

Textbook Reading

- This lab appears on page **290** of the *Engineering Our Digital Future* textbook.
- Prerequisite textbook reading before completing this lab: pp. **281-289**.

Engineering Designs and Resources

Worksheet used in this lab:

- **L05-02-01 Quantization and Clipping.Lst**: Worksheet demonstrating the effects of using too few bits for quantization and the clipping effect.

Worksheet Description

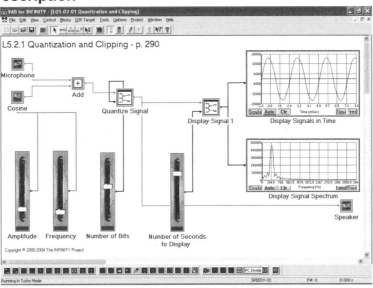

Figure 5.4 Quantization and Clipping

The input signal is the sum of the microphone and a cosine. This signal is then quantized and the result is displayed both as a signal vs. time (top) and as frequency (bottom).

The input of the microphone is added to a Cosine signal which is quantized, display (both time and frequency waveforms) and emitted from the speaker. You can adjust the amplitude and frequency of the Cosine, the number of bits to which to quantize the signal and the number of seconds to display in the waveform plots.

Open the worksheet *L05-02-01 Quantization and Clipping.Lst.* Right click the number of bits slider. Change the number of steps from 101 to 16. Then click **OK**.

You should see something similar to the Figure 5.4 above except the plots will be empty.

Laboratory Procedures

Steps	Instructions
Step 1:	Start the worksheet.
Step 2:	Set the amplitude slider to 20,000. Set the frequency slider to 413 Hz. Increase the Amplitude slider to achieve a low but audible tone, and set the Number of Bits slider to its maximum value.
Step 3:	Slowly decrease the Number of Bits slider until the tone begins to change.
	Q1: What is the number of bits at which this first happens? What changes do you hear in the sound as the number of bits is decreased to 1?
	Q2: Is the noise in the signal noticeable in the time plot (top)? What about in the frequency plot (bottom)?
	Q3: Sketch each plot for high values of Number of Bits (per sample) and low values.
Step 4:	Set the amplitude slider to 5,000 and repeat Step 3. You may want to increase the volume on the speaker to hear this signal better.
	Q4: What is the number of bits at which this first happens? What changes do you hear in the sound as the number of bits is decreased to 1?

Steps	Instructions (Continued)
	Q5: Is the noise in the signal noticeable in the time plot (top)? What about in the frequency plot (bottom)?
Step 5:	With the amplitude set at 5,000, set the number of bits per sample to 4.
	Q6: Count the number of different levels in the time signal plot. How many possible levels could there be? Why aren't all of them used? (This will be easier if you stop the worksheet to examine the plot and then restart it.)
Step 6:	With the amplitude set at 5,000, set the number of bits per sample to 3.
	Q7: Count the number of different levels in the time signal plot. How many possible levels could there be? Increase the amplitude to 10,000. Now how many different output levels are used? Why are more used for the higher amplitude signal?
	Q8: Based on the answer to the previous question, what amplitude do you think would use all the possible levels?
Step 7:	Reduce the Amplitude slider to zero. Now only the microphone is being quantized and displayed. Reset the number of bits to 16.
Step 8:	Have you or your lab partner speak into the microphone, while you reduce the Number of Bits slider until you hear noise?

Steps	Instructions (Continued)
	Q9: At what value of bits/sample do you hear noise?
Step 9:	Continue to reduce the Number of Bits slider while talking into the microphone.
	Q10: At what point can you no longer tell who is speaking?
	Q11: At what point can you no longer understand what is being said?
Step 10:	Stop the worksheet
Step 11:	Turn the speaker all the way down. We will no longer hear the output of the worksheet. We are going to make a very large volume signal.
Step 12:	Increase the amplitude up to 35,000.
	Q12: What happens to the display of the time and frequency plots? Stop the worksheet to look closely at the plots, then restart the worksheet.
Step 13:	Watch the frequency plot as you increase the amplitude from 35,000 to 60,000.
	Q13: Describe what happens to the time and frequency plots.
Step 14:	Increase the amplitude from 60,000 to 100,000.

Steps	Instructions (Continued)
	Q14: Describe what happens to the time and frequency plots. How is this display similar to Lab 5.1.2?
Step 15:	Turn up the volume on the speaker just a little so that you can comfortably hear the sound.
	Q15: What changes do you hear in the sound as you reduce the amplitude to 40,000? How is the sound related to the frequency plot?
	Q16: What do you observe? Sketch the waveform in the time display.
	Q17: Based on your observation, estimate the maximum value the time waveform can achieve. This effect is called clipping. How does this compare to your answer to Question 8?

Overview Questions

A: What is quantization noise?

B: What is a good and useful measure of the effort of quantization noise?

Summary

When a limited number of bits is used to represent a number once it is sampled in a computer, there can be errors in the waveform due to quantization and clipping.

Coding Information for Storage and Secrecy

Chapter 6 focuses on coding information so it can be stored on a computer using only bits. This is important for computer security and encryption. Redundancy of numbers makes data compression possible. The basic concepts of detecting and correcting errors in digital data are discussed. This chapter also gives students some very interesting applications of simple polynomials and random numbers.

Infinity Labs

6.1 Speech Compression
6.2 Rotational Encoder and Decoder
6.3 Pseudo-Random Number Generator

Introduction

The laboratories in Chapter 6 focus on the encoding and encryption of data. Lab 6.1 focuses on the relatively little amount of information contained in a voice signal as compared to its uncompressed data rate. This results in a dramatic ability to compress the voice data with negligible loss of information. Lab 6.2 centers on the use of a simple rotational encoder. It will allow you to encode, transmit, and decode messages with a single key. Finally, Lab 6.3 allows you to inspect the functionality of the underlying "random" number generation of encryption and other computer codes. By manipulating the coefficients of an iterative mathematical equation you can create strings of "random" and not so random numbers.

The basic input blocks of Chapter 6 include audio file read blocks and keyboard input blocks. The output blocks are speakers, and graphical displays. In this Chapter you can examine the effects of compression on voice data and interact with the mathematical structures underlying encryption algorithms.

6.1 Speech Compression

Lab Objectives

Sounds can be compressed to reduce the amount of space needed to store them or to reduce the effort it takes to transmit them. The sound files can be made dramatically smaller–up to 1/30th of their original size, for example. The sound can be changed somewhat, however. This lab contains examples of lossy compression, in that some information is lost in the compressed signals. How important is the lost information? Let your ears be the judge.

Textbook Reading

*   This lab appears on page **330** of the *Engineering Our Digital Future* textbook.
*   Prerequisite textbook reading before completing this lab: pp. **305-329**.

Engineering Designs and Resources

Worksheet used in this lab:

*   **L06-01-01 Speech Compression.Lst**: Students study the effects of speech compression on the intelligibilty of sounds.

This lab demonstrates the effect of compressing a particular speech signal to use less space to store it or less effort to transmit it. The original speech signal was recorded using 16 bits per sample and 8,000 samples per second, so it requires 128,000 bits or 128 kbits to store each second of speech. If the storage were reduced by a factor of 4, we would use only 32 kbits to store a second of speech.

The sound files are located in the "C:\Program Files\Hyperception\VABINF\Sounds\" folder. Each sound file contains a male and female voice repeating the same statement.

Female: "The pipe began to rust while new."

Male: "Oak is strong and also gives shade."

These sentences were selected to give a wide variety of sounds found in speech. The various compressed sound files have been recorded with differing compression standards offering differing amounts of compression. The table at the top of the next page lists the sound files, bit rates, and compression standards used.

Many different methods may be used to compress speech because some applications can tolerate more loss of speech quality due to compression than other applications. The specific methods used in this laboratory are shown in the table on the next page. The names of the compression techniques may include an abbreviation which tells you the name of the method. For example, ADPCM stands for adaptive differential pulse code modulation, which was developed in the 1940s. The names may also include the name of an accepted national or international standard, such as G.728, that completely describes all the details of how the compression is done and how the compressed speech is restored so you can listen to it. Standards are necessary so that many users in different places can use each other's compressed data files.

Sound File Name	Kbits/sec data rate	Compression Standard
32.wav	32 kbits/sec	G.726 ADPCM 32kbps (Cordless phone) standard
16.wav	16 kbits/sec	G.728 LD-CELP standard
8.wav	8 kbits/sec	IS-54 US Cellular (VSELP) standard
4.wav	4 kbits/sec	FS 1016 US DoD CELP standard
2.wav	2 kbits/sec	MELP US DoD Mixed CELP standard

Worksheet Description

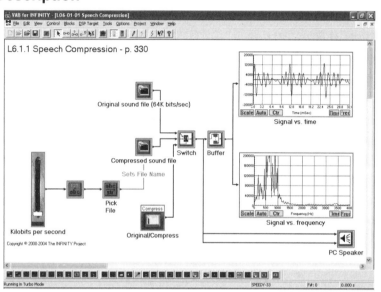

Figure 6.1 Speech Compression

Open the worksheet *L06-01-01 Speech Compression.Lst*. You should see something similar to the Figure 6.1 above.

The slider on the lower left ("Kilobits per second") can take on the values 2, 4, 8, 16 and 32. It will be used to select a compressed sound file to hear (the one which has the matching name). This value is used to create a file name corresponding to a sound signal sampled with that compression rate by the next two blocks. The next block is a switch which allows either the compressed file to be played and displayed OR allows the corresponding uncompressed file to be played and displayed in its place. Which file the switch chooses is controlled by the push button at the bottom of the screen labeled "Original/Compressed".

Laboratory Procedures

Steps	Instructions
Step 1:	Start the worksheet.
Step 2:	Be sure the **Original/Compress** button shows the text "Original". If you stop the worksheet and restart it, you may have to push this button twice to make sure it is correct.
Step 3:	Listen to the file carefully. The differences in the compressed files will be small.
Step 4:	Set the "Kilobit per second" slider to 32 and press the **Original/Compressed** worksheet button. Note that the bottom title says "Compress". You are now hearing the 32 kbit/sec version of the sound file.
Step 5:	Click the **Original/Compressed** worksheet button several times to alternate listening to the compressed and original files.
	Q1: Can you hear any difference?
Step 6:	Move the Kilobit per Second slider down to 16.
Step 7:	Click the **Original/Compressed** worksheet button several times to alternate listening to the compressed and original files.
	Q2: Can you hear any difference now?
Step 8:	Continue to decrease the Kilobits/sec slider (thereby increasing the compression) and listen to the compressed and original versions.
	Q3: At what value of kilobits/sec do the two signals begin to sound different.

Steps	Instructions (Continued)
	Q4: Can you still understand the statement at 2 kbits/sec?
Step 9:	At 2 kilobits per second, observe the signal vs. frequency plot as you switch between compressed and uncompressed speech. Pay particular attention to the 2,500 Hz to 3,500 Hz region when the female speaker is saying words with a "p" in them.
	Q5: What is the main difference between the two speech signals in this frequency range?
Step 10:	In Lab 5.1.3 you created compressed speech by the very simple method of keeping only one sample out of every four that came from the microphone. Assuming that the samples from the microphone have 8 bits each this would be a 16 kbit/s speech signal. Compare the effects on your speech in the Lab 5.1.3 worksheet with the effects of the 16kbit/s compression on the stored speech in this laboratory.
	Q6: Which compression method do you think is most efficient at preserving the important information in the speech signal?
	Q7: A sound file is recorded with a sample rate of 8,000 Hz and a quantization of 16 bits per sample. What is the data rate in the sound file? If this file were compressed to achieve the data rates used in the lab, what would that make the compression ratio for each of the sound files in this lab?

Steps	Instructions (Continued)
	Q8: Suppose each sound file in this lab has a sample rate of 8000Hz. How many bits/sample are used for each of the compressed sound files?
Step 11:	Stop the worksheet Close the worksheet

Summary

There is quite a bit more information in the human voice than we need just to understand speech. As you have seen from this lab, we can greatly reduce the number of bits in a speech signal and retain good quality.

6.2 Rotational Encoder and Decoder

Lab Objectives
Making messages secret is an important business—all electronic commerce and banking depends on it. In this lab, we look at a very simple encryption method that was used back in the days of the Greeks and Romans—and today, it is often used as an entertaining puzzle in the Sunday newspaper.

Textbook Reading
- This lab appears on page **342** of the *Engineering Our Digital Future* textbook.
- Prerequisite textbook reading before completing this lab: pp. **337-340.**

Engineering Designs and Resources
Worksheets used in this lab:

- **L06-02-01 Rotational Encoder and Decoder.Lst** : A simple rotational encoder, that maps one character to another through a simple rotation of the entire alphabet.
- **L06-02-02 Rotational Encoder and Decoder Transmit.Lst**: Students get to transmit encrypted messages over the internet.
- **L06-02-03 Rotational Encoder and Decoder Receive.Lst**: Students receive and decrypt messages sent by the transmitter.

6.2.1 Rotational Encoder and Decoder

Worksheet Description

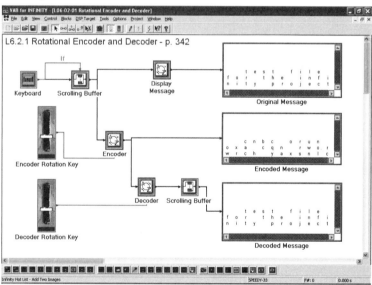

Figure 6.2 Rotational Encoder and Decoder

Open Worksheet *L06-02-01 Rotational Encoder and Decoder.Lst*. You should see something similar to Figure 6.2.

A simple way of encoding a message is to substitute one letter for another. For example the following list maps A to B, and B to C, and so on to Y to Z and Z to A.

Input Letter: A B C D E F G H I J K L M N O P Q R S T U V W X Y Z

Output Letter: B C D E F G H I J K L M N O P Q R S T U V W X Y Z A

This encoding rotated the alphabet 1 letter, so it's key is 1.

This worksheet performs rotational encoding and decoding of a message typed on the keyboard. In this lab we will experiment with this simple rotational encoder/decoder.

Laboratory Procedures

Steps	Instructions
Step 1:	Start the worksheet.
Step 2:	Adjust the Encoder Rotation Key Slider to zero (0) and the Decoder Rotation Key Slider to zero (0).
Step 3:	Type a short message on your keyboard. It should have appeared in all three text displays on the right hand side.
Step 4:	Now adjust the Encoder Rotational Key slider to a value between 1 and 25. The second text display shows the encoded message. The third text display shows the decoded message, which in this case was improperly decoded since we left the Decoder Rotational Key set to zero (0).
Step 5:	Look at the first letter of your encoded message and slowly increase the value of the encoding key from 0 to 26.
	Q9: With an Encoder key value of 17, to what letter does the letter A map?
	Q10: Explain how the encoded letter changes as you increase the encoding key. What happens when the key is 26?
Step 6:	Select an encoder key that is neither 0 nor 26.
	Now adjust the Decoder Rotational Key to be the same value as the Encoder Rotational Key. The decoder works exactly the same as the encoder, except that it just rotates the alphabet in the opposite direction.

Steps	Instructions (Continued)
	Q11: As you adjust the Decoder Rotational Key to other values, do any of the incorrect values (not matching the encoder key) result in a readable message? Why or why not?
	Q12: Decode the following messages: QVNQVQBG with key = 8 Zai ue ftq fuyq with key = 12 Hxd pxc rc (Hint: if you set the Encoder Key to 0, it will decode what you type in.) If you got the last message you now know how easy it is to crack a message coded this way.
Step 7:	Stop the worksheet. Close the worksheet.

6.2.2   Rotational Encoder and Decoder - Transmit

Worksheet Description

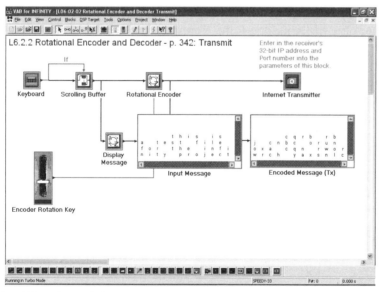

Figure 6.3 Rotational Encoder and Decoder - Transmit

Open the worksheet *L06-02-02 Rotational Encoder and Decoder Transmit.Lst.* You should see something similar to Figure 6.3.

If you have network access and a partner you can send the encrypted message over the network. The person who is sending the message will need to use worksheet *L06-02-02 Rotational Encoder and Decoder Transmit.Lst*, whereas the person receiving the message will need to use the worksheet *L06-02-03 Rotational Encoder and Decoder Receive.Lst.* You will need to know your IP (network) address to use these sheets. To determine your IP address from the windows operating system:

*   Click **Start**, select **Run**
*   Type **cmd** and click **OK** to bring up a command window
*   In the command window type **ipconfig**. Record the ip address of your computer
*   Type **exit** in the command window to close it

You will also need to agree on a port number to use. The default port number for these worksheets is 6000, use it unless you have problems.

Laboratory Procedures

Steps	Instructions
Step 1:	BEFORE YOU CLICK **GO**: You will need to enter the IP address of your lab partner's computer in this worksheet. If you should click **Go** before doing this the computer will attempt to make a connection to an unspecified computer and will appear to hang until the connection times out.
Step 2:	Double click the **Internet Transmitter** Block. This will open up a dialog box to allow you to enter the IP address to which you wish to transmit (i.e., your lab partners) as well allowing you to change the port number if required. Enter your lab partner's IP address and click **OK** to close the dialog box.
Step 3:	At this point, wait until your lab partner has reached Step 3 of their directions and clicked **Go** on their sheet.
Step 4:	Click **Go**.
Step 5:	Select an Encoder Key Value and begin typing. The encoded message is being transmitted to your lab partner. See if they can figure out the key without your telling it to them. [No fair typing nonsense! You must type a real message if they are to guess the key.]
Step 6:	Stop the worksheet. Close the worksheet.

6.2.3 Rotational Encoder and Decoder - Receive

Worksheet Description

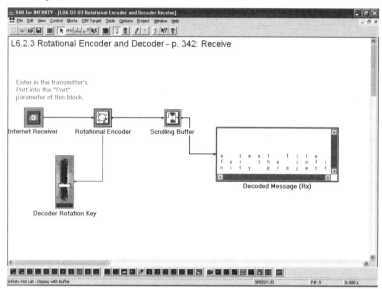

Figure 6.4 Rotation Encoder and Decoder - Receive

Open the worksheet *L06-02-03 Rotational Encoder and Decoder Receive.Lst*. You should see something similar to Figure 6.4.

This worksheet will allow you to receive and decode the text messages your lab partner sends you.

Laboratory Procedures

Steps	Instructions
Step 7:	BEFORE YOU CLICK **GO**: You will need to enter the port number that you and your partner have agreed to use.
Step 8:	Double Click the **Internet Receiver Block.** This will open up a dialog box to allow you access to the port number. Enter the port number you have agreed on and click **OK** to close the dialog box.
Step 9:	Click **Go** to begin waiting for messages.
Step 10:	You should see text appear as your lab partner begins to send you encoded data.
	Q1: Can you figure out the key without being told? How did you do it?

Overview Questions

A: How could we make this encoding method more secure while keeping a relatively simple structure?

Summary

This type of coding is a simple to implement but relatively easy to crack.

6.3 Pseudo-Random Number Generator

Lab Objectives

Have you ever wondered how your computer generates "random" numbers? As it turns out, the numbers a computer generates are not really random at all. The computer generates them using a "predictable" formula. If you don't know the formula the numbers look awfully random. That's why this approach is called pseudo-random number generation.

Textbook Reading

• This lab appears on page **349** of the *Engineering Our Digital Future* textbook.

• Prerequisite textbook reading before completing this lab: pp. **346-348.**

Engineering Designs and Resources

Worksheet used in this lab:

• **L06-03-01 Pseudo-Random-Number Generator.Lst**: Students experiment with a simple pseudo random number generator.

Worksheet Description

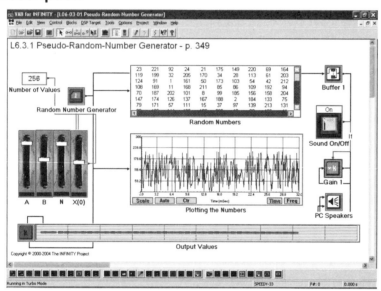

Figure 6.5 Pseudo-Random Number Generator

Open the worksheet *L06-03-01 Pseudo-Random-Number Generator.Lst*. You should something similar to Figure 6.5.

This worksheet generates lists of pseudo-random numbers as described in the book. It uses slider values to set the values of A, B, N and X(0), the seed value. Additionally, you can set the number of values to display in a numeric entry box in the upper left corner of the worksheet.

The output numbers are computed using X(n+1) = [A*(n) + B] mod (N) from page 348 of the text. The initial value, X(0), is called the seed value.

Laboratory Procedures

Steps	Instructions
Step 1:	Start the worksheet.
Step 2:	Observe the random numbers generated.
Step 3:	Click the **Sound On/Off** button to hear a waveform created by the random numbers.
	Q1: What do the random numbers sound like?
Step 4:	The Output Value display plots the values contained in the repeating list of "random" numbers. The more sparse it appears, the shorter the sequence is before repeating. Experiment with the values of A, B, N and X[0] to find a sequence that repeats itself after only a few numbers (i.e., there are only a few green dots plotted in the display. Hit **Esc**, or Click **Stop** to pause the worksheet.
	Q2: Scroll through the list of random numbers. How many numbers are there before the first number is repeated?
	Q3: Where does the second number in the list repeat as compared to the first number?
Step 5:	Using the sliders, set A=472, B=131, N=225. Click **Go**. Now slide the seed value (X[0]) to various values. (You may wish to make the size of the X[0] slider larger so you can choose any value.)
	Q4: Based on the density of the green dots, what are examples of good choices for X[0]? What are some bad examples?

Steps	Instructions (Continued)
	Q5: What is the smallest number of green dots you can make? Can you get it down to 9?
Step 6:	Using the sliders, set A=742, B=103, and N=193.
	Q6: Can you find a good value for X[0]? Do most of the values of X[0] produce the same set of green dots? Can you find a bad value? If so, what is it?
Step 7:	Repeat Steps 5 and 6 with the speaker turned on.
	Q7: Describe the sound you hear out of the speaker when you have selected "good" settings for the random number generator versus the sound when you have selected "bad" settings.
Step 8:	Using the sliders, set A=243, B=159, N=121, and X[0]=29.
	Q8: Examining the plot of output values, are most of the numbers contained in the sequence?
	Q9: How does the sequence look on the time display? (You might want to stop the sheet to take a closer look.)

Steps	Instructions (Continued)
Step 9:	Click the **Freq** button on the Plot to examine the frequency content of the random number list.
	Q10: Is the frequency plot relatively flat or do some spikes grow taller than others? (Note: All of these plots will have a high value (a "spike") at 0 Hz as all of the numbers we are generating are positive. You can ignore that spike.)
	Q11: Based on your answers to Questions 8, 9 and 10, is this a good choice of settings for a random number generator?
Step 10:	Try other values of A, B, N and X[0] to find one you think works well and one that works poorly.
	Q12: What are your "Good" settings and your "Bad" settings?
	Q13: Describe the Output Values plot for each. **Q14:**
	Q15: Describe the Sound from the speaker for each.
	Q16: Describe the "Plotting the Numbers" display for each.
	Q17: Describe the "Plotting the Numbers" display of frequency for each.

Steps	Instructions (Continued)
Step 11:	Stop the worksheet. Close the worksheet.

Optional Quantitative Investigations: How do A, B, and N affect the sequence of numbers? When N is a small number, such as 7, it is easy to investigate the effects of different values of A and B.

Steps	Instructions
Step 1:	Open the worksheet *L06-03-01 Pseudo-Random-Number Generator.Lst.*
Step 2:	Start the worksheet.
Step 3:	Turn the sound off. Set A=1, B=1, N=7, and X(0) = 1.
Step 4:	Stop the worksheet.
	Q1: What sequence of numbers do you see in the random number display? Verify that you should see this sequence using $X(n) = (A*X(n-1) + B) Mod(N)$. For N=7, what is the longest pattern we could see before the pattern would start to repeat?
	Q2: Is the Plotting Numbers display consistent with your answer to Question 1?
	Q3: Is the Range of Output Values consistent with your answer to Question 1?
	Q4: What happens if you change X(0) to another number and restart the worksheet? Why?
Step 5:	Set X(0) back to 1 and set B to 3. Restart the worksheet. Then stop the worksheet to observe the results.

Steps	Instructions (Continued)
	Q5: What sequence of numbers do you see in the random number display? Verify that you should see this sequence using $X(n) = (A*X(n-1) + B) \, Mod(N)$.
	Q6: Is the Plotting Numbers display consistent with your answer to Question 5?
	Q7: Is the Range of Output Values consistent with your answer to Question 5?
	Q8: What happens if you change $X(0)$ to another number and restart the worksheet? Why?
Step 6:	Set $X(0)$ back to 1. Restart the worksheet. Turn the sound on and switch B between values of 1 and 3.
	Q9: How does the sound change? Why?
Step 7:	Turn the sound off and stop the worksheet.
Step 8:	Change the top value of A to 254 by clicking the **A** slide and setting top = 254. Set A=3, B=1, N=7, and $X(0) = 1$. Start the worksheet and then stop it to observe the results.

Steps	Instructions (Continued)
	Q10: What sequence of numbers do you see in the random number display? Verify that you should see this sequence using X(n) = (A*X(n-1) + B) Mod(N).
	Q11: Why does this sequence repeat after a shorter number of values than the sequences in Steps 3 and 4?
	Q12: What number is missing from the repeating sequence observed in Question 10. Verify this by observing the Output Values display.
Step 9:	Restart the worksheet. Set X(0) to the missing value from Question 12.
	Q13: Explain what happens on the worksheet. Verify that this should happen using X(n) = (A*X(n-1) + B) Mod(N).
Step 10:	Set B=5 and keep A=3, N=7, and X(0) = 3.
	Q14: Now which value is missing from the sequence? Look at the Output Values display to see this.
Step 11:	Adjust X(0) to this value and verify that the output stays at this value.

Steps	Instructions (Continued)
Step 12:	For A=3 and N=7, use the worksheet find the sequences and the missing values in the sequence for all seven values of B. Set the top values of the sliders for B and X(0) to 20 and the bottom values to 0 to make adjustments easier.
	Q15: $\underline{B}$ <u>sequence</u> <u>missing value</u> 0 1 2 3 4 5 6
Step 13:	Set A=3, B=1, N=16, and X(0) = 0. Observe the Output Values display as you adjust X(0).
	Q16: How long is the output sequence? How does it change when X(0) changes?
Step 14:	Change B to a value of 10 and repeat Step 12.
	Q17: Does the length of the output sequence depend on X(0)? How many different lengths do you observe?
Step 15:	Stop your Worksheet. Close the worksheet.

Overview Questions

A: What happens as you display more and more values in the pseudo-random number sequence?

B: Give an example of a practical system where pseudo-random numbers are used.

Summary

The "random" numbers a computer generates are not very random, are they? The makings of a random number generator that seem random depends greatly on a good choice of A, B, and N.

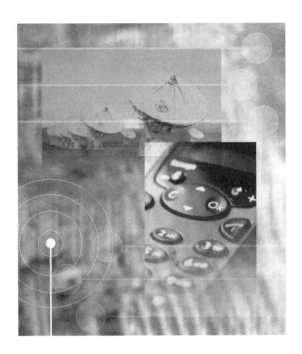

Communicating with Ones and Zeros

In Chapter 7 we introduce the design of an "air modem" which uses audio signals to transmit information. In the chapter we make successive iterative improvements to the design to make it simpler and more robust. Here we will build several "air modems" and examine their characteristics.

Infinity Labs

Introduction

The laboratories in Chapter 7 center on the use of audio signals to transmit information. An incrementally improving approach is presented in which audio frequencies are used to transmit text based messages with increasing efficiency and decreasing complexity. The continual improvement of the design in terms of the number of audio frequencies required and the corresponding design complexity results in designs analogous to standards used today: frequency shift keying, touch tone telephone operation, and fax machines transmissions.

The basic input blocks of Chapter 7 are the keyboard text input blocks and image read and creation blocks. The input data are converted into audio frequencies through the use of Sine and Cosine generators which are "played" over the speaker for data transmission.

7.1 Audio Communication of Messages Using One Tone Per Letter

Lab Objectives

In these labs we will construct a simple air modem. For this initial design, each character has a unique audio frequency. Figure 7.1 shows the concept. The message to be transmitted t is examined on a character by character basis. Each character is converted into a burst of a single frequency based on a table. This frequency is then played over a speaker, where it is sampled by the receiver and its frequency is determined, thereby identifying the character that was transmitted.

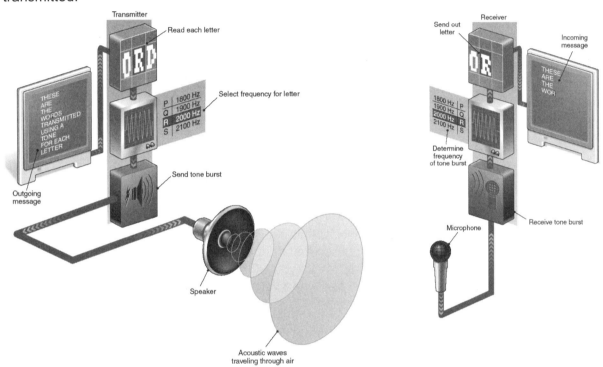

Figure 7.1 Simple air modem in which each character is converted into a single audio frequency by the transmitter and converted back into a character by the receiver.

Textbook Reading

- This lab appears on page **365** of the *Engineering Our Digital Future* textbook.
- Prerequisite textbook reading before completing this lab: pp. **361-365.**

Engineering Designs and Resources

Worksheets used in this lab:

- **L07-01-01 Audio Communication Using One Tone Transmit.Lst**: Transmit messages using a burst of a single frequency per character.
- **L07-01-02 Audio Communication Using One Tone Receive.Lst**: Receive messages sent by lab procedures in worksheet L7.1.1.

7.1.1   Audio Communication Using One Tone Transmit

Worksheet Description

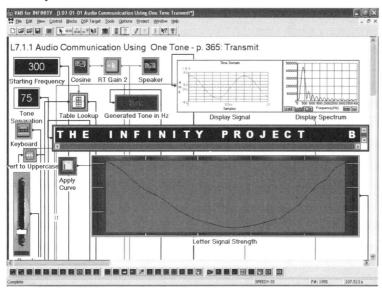

Figure 7.2 Audio Communication Using One Tone Transmit

Open the worksheet *L07-01-01 Audio Communication Using One Tone Transmit.Lst*. You will see something similar to Figure 7.2.

This worksheet is an implementation of the left half of Figure 7.1 above. As this sheet is more complex than can be seen on the computer screen at one time, much of the processing is performed off of the bottom of the screen. The blocks you will interact with, either by entering data or examining results are all visible on the screen when you first open it.

In Figure 7.1 the input to the transmitter is provided by a text message. In the worksheet this is provided by the keyboard block on the far left-hand side of the sheet. The table lookup function is performed by three blocks on the worksheet: Table Lookup, Starting Frequency numeric entry, and Tone Separation numeric entry. Using these blocks, a letter you type on the keyboard determines a unique frequency. The Table Lookup block converts the letter into an ordinal value (i.e., A=1, B=2, Z=26) and the two numeric entries assign a frequency. The first ordinal letter (A) is assigned the Starting Frequency, and each following letter is assigned a frequency one "Tone Separation" higher than the previous. So if the "Starting Frequency" is 300 Hz and the "Tone Separation" is 75 Hz then A = 300 Hz, B = 375 Hz, C = 450 Hz.

The frequency value of the current character is displayed in the "Generated Tone in Hz" numeric display. The duration of the tone burst is set by the Slider at the lower left-hand side of the worksheet. The message you are typing along with the time-domain and frequency-domain plots of the tone burst are displayed for you. Finally, the "Letter Signal Strength" graphical data entry and the "Apply Curve" button are used to calibrate the air modem. We will describe this graph in detail later. For now, let's interact with just the transmitter worksheet.

Laboratory Procedures

Steps	Instructions
Step 1:	Start the worksheet.
Step 2:	Click the **Apply Curve** button to initialize it.
Step 3:	Slowly type your name.
	Q1: Could you distinguish the different audio frequencies for each letter?
Step 4:	Alternate between typing your name and your lab partner's name.
	Q2: How easy is it to differentiate between two different words once you have heard them several times?
	Q3: Can you tell if you mistyped without looking at the text output?
Step 5:	Examine the spectrum plot as you type.
	Q4: Describe the spectrum of a single letter.
	Q5: Identify the peak of the spectrum for any letter.

Steps	Instructions (Continued)
	Q6: Does this peak correspond to the value of the "generated Tone in Hz" numeric display?
Step 6:	Change the "Burst Length" to a longer value by adjusting the slider.
	Q7: How did the sound change?
	Q8: How did the spectrum display change?
	Q9: Explain your answers to Questions 7 and 8.
Step 7:	Reset the burst length to 0.35. Now change the tone spacing by typing 10 in place of 75 and hitting enter. Try typing your name again. Also try typing letters that are closely spaced like "jklmn".
	Q10: Can you still hear the difference between different characters? Is it easier or harder?
Step 8:	Change the tone spacing to 250.

Steps	Instructions (Continued)
	Q11: How easy is it to differentiate between different characters frequencies now?
Step 9:	Type the alphabet. Pay close attention to what happens between the letters P and Q.
	Q12: Does the displayed value of "Generated Tone in Hz" match the peak of the spectrum?

Now you have an appreciation for the fact that longer tones give more precise frequencies, but reduce the number of characters per second transmitted. You have also noted that increased tone separation makes differentiating frequencies easier, but at the expense of using more of the available audio frequency range. Next we will look at the receiver worksheet and get the two to work together to form a simple communications link.

7.1.2 Audio Communication Using One Tone; Receive Side

Worksheet Description

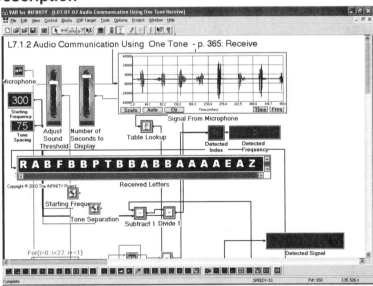

Figure 7.3 Audio Communication Using One Tone Receive

Open the worksheet *L07-01-02 Audio Communication Using One Tone Receive.Lst*. You will see something similar to Figure 7.3.

The basic Air Modem Receiver is depicted in the right hand side of Figure 7.1. At the heart of the Air Modem receiver is a filter bank, which allows the receiver to determine the presence or absence of any known frequency in the signal. Figure 7.4 is an example of such a filter bank. In this simple Air Modem, each character is represented by a single frequency so only one filter in the bank should respond with a large signal.

In the worksheet depicted in Figure 7.3, much of the complexity, again, does not fit on the screen. The microphone samples the audio signal. A slider adjusts the threshold for declaring that a received signal is one of the frequencies of interest. If such a threshold didn't exist, then, when no signal was being transmitted, the filter in the bank with the largest (albeit small) value would declare that "silence" as its letter. The detected frequency is displayed numerically along with a time based representation of the received waveform. Next the "starting frequency" and "Tone spacing" numeric entries are used to determine which ordinal character was transmitted. As this process represents the inverse of the transmitter's conversion to a frequency, it is important that the Starting Frequencies and Tone Spacing settings match for a transmitter / receiver pair. Finally the determined character is printed in a text display entitled "Received Letters".

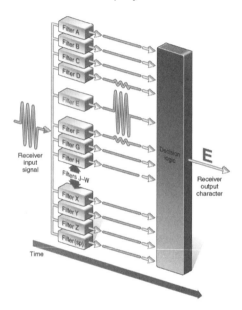

Figure 7.4 Filter Bank for the Simple Air Modem. Since only one frequency is present in the transmitted signal, only one filter should have a strong response

Laboratory Procedures

Steps	Instructions
Step 1:	Team up with a lab partner. One of you will run the transmitter worksheet *L07-01-01 Audio Communication Using One Tone Transmit.Lst* and the other will run this receiver worksheet *L07-01-02 Audio Communication Using One Tone Receive.Lst*
Step 2:	Place the microphone of the "receiver" DSP kit very near the speakers of the "transmitter".
Step 3:	Make sure you have selected the same Starting Frequency and Tone Separation used at the transmitter.

Steps	Instructions (Continued)
Step 4:	Open the worksheets listed in Step 1. Start the worksheets.
Step 5:	Transmitter: Slowly type the alphabet
	Q1: Where any of the letters correctly received by the receiver?
Step 6:	If not, then increase the speaker volume of the transmitter and/or decrease the "Adjust Sound Threshold" until several letters are correctly received.
Chances are that not all letters were correctly received. Let's examine why:	
Step 7:	Stop the receiver worksheet during the reception of a letter that is being received and recognized.
	Q2: What is the maximum amplitude of the frequency burst on the receiver's "Signal from Microphone"?
Step 8:	Start the receiver again and halt it during the reception of a letter that is not being recognized properly.
	Q3: What is the maximum amplitude of the frequency burst on the receiver's "Signal from Microphone"?
In all likelihood, the character that was not recognized had a lower received amplitude than the one that was recognized properly. Yet the transmitter worksheet is attempting to play all the frequency bursts with the same amplitude (look at the "Display Signal" plot of the Transmitter worksheet to verify this).	
	Q4: Where does the signal pass after it is generated electronically on the transmitter and before it is displayed on the receiver worksheet? Which of these components might be responsible for reducing the signal amplitude in a frequency dependent way?

We've built a mechanism for adjusting for these unpredictable frequency dependent volume changes into the transmitter worksheet. This process is referred to as Calibrating Your Worksheet.

Calibrating Your Worksheet

One of the first steps when you opened your transmitter worksheet was to click the **Apply Curve** button. That action applied the line evident in the "Letter Signal Strength" graphical entry box to independently adjust the volume of each character. The default line was a straight horizontal line. Therefore, up until now, every character is played with the same volume. We are going to adjust this to compensate for the fact that somewhere later in the system certain characters are not being received as loudly. You will need to work in pairs to calibrate your link. If you physically move the system, you may have to recalibrate it.

1. Start with a letter that works well. Play this letter on the transmitter and examine its amplitude on the "Signal from Microphone" display on the receiver. Record this value. You will attempt to make every letter be received just a well as this one.

2. The letter signal strength graphical data entry on the transmitter has 27 points along its horizontal axis corresponding to the 26 letters plus the space character. If you click

your mouse with your cursor within this graphical data entry block you adjust the values of the yellow line. The value of the first data point represents the relative volume of the letter A. The next point represents the value for the letter B, etc., all the way through the last value which represents the relative volume for the space. Adjust the volume of the letter A by changing the left most portion of the yellow curve and clicking the **Apply Curve** button (no change takes effect until you click **Apply Curve**) until it is loud enough to be correctly received at the receiver.

3. Next move on to the letter B, adjusting the next portion of the yellow curve.

4. As you move through the alphabet in this manner you will likely find there are many letters in the middle which are required to be louder than the rest to be received correctly. A typical calibration curve is depicted in Figure 7.5.

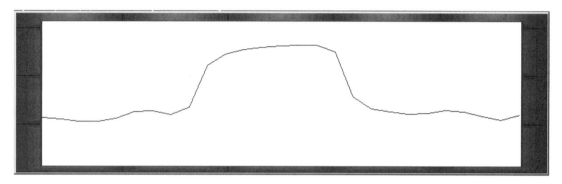

Figure 7.5 Typical calibration curve for Air Modem

5. If you can't reliably receive all the characters you can try adjusting the transmitter speaker volume louder, the receiver "adjust sound threshold" slider lower and the tone duration longer. Also you can try increasing the Tone spacing on both the transmitter and receiver.

6. Once you have a reliable communications link move on to the next steps.

Next Steps

Steps	Instructions
Step 1:	Now that you've established a reliable communications link, let's try some modifications to it. On the receiver only, adjust the tone spacing to half the value of the transmitter tone spacing. Type a message.
	Q1: Did you receive a letter for each letter sent?

Steps	Instructions (Continued)
	Q2: Were they correct?
	Q3: Can you discern the mapping between the letters that were transmitted and those received?
	Q4: Could you transmit a message this way? What letters could it contain?
Step 2:	Repeat this experiment with the receiver "Tone Spacing" set to twice the value of the transmitter.
	Q5: Did you receive a letter for each letter sent?
	Q6: Were the ones your received correct?

Steps	Instructions (Continued)
	Q7: Can you discern the mapping between the letters that were transmitted and those received?
	Q8: Could you transmit a message this way? What letters could it contain?

7.2 Effects of Weak Signals and Noise

Lab Objectives
Same as Lab 7.1 objectives.

Textbook Reading
- The lab appears on page **377** on the *Engineering Our Digital Future* textbook.
- Prerequisite textbook reading before completing this lab: pp. **376-377**.

Engineering Designs and Resources
Worksheets used in this lab:

- **L07-02-01 Effects of Weak Signals and Noise - Transmit**: Allows you to add noise to the transmission worksheet of Lab 7.1 to analyze its effects.

Worksheet Description

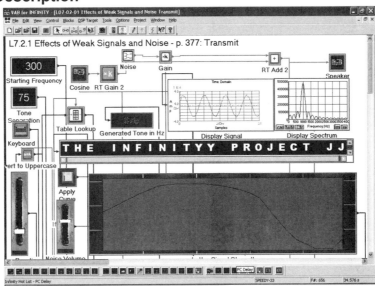

Figure 7.6 Effects of Weak Signals and Noise - Transmit

Now, unless you are in a quiet room, you've probably encountered noise in your system already. You can increase the noise in your system relative to the signal strength of the communications link by several methods:

- increasing the distance between the speakers and microphone,
- putting something between the speakers and microphone (the item could vary from fabric to a book),
- to talking loudly near the microphone as the communication happens.

If you don't have any real noise handy, you can **open** the transmitter worksheet *L07-02-01 Effects of Weak Signals and Noise.Lst*.

This worksheet is identical to the Transmitter worksheet you've been using with the exception that it has a noise slider near the lower left-hand corner which adds noise to the signal before it is transmitted. The goal of this slider is to mimic what would have happened had there been additive noise in between the transmitter and the receiver.

Steps	Instructions
Step 3:	Introduce noise into the communications link by one of the methods suggested above.
	Q9: What is the impact on your system?
	Q10: Can you increase the speaker volume to overcome the noise?
Step 4:	Measure the peak amplitude of the received characters on the receiver worksheet.
Step 5:	Increase the noise source until it just begins to stop working and measure the amplitude of the noise on the receiver worksheet (it will be most evident between the bursts or when the transmitter is silent).
	Q11: What is the ratio of the signal to the noise when you system just stops working?
	Q12: Write down the frequencies used by both teams to send the letters A B and C below:

Letter	Team #1 Frequency	Team #2 Frequency
A		
B		
C		

Q13: Do they seem to be interfering?

Q14: At what point do they begin to interfere?

7.3 Different Codes

Laboratory Objectives
Same as objectives in Lab 7.1.

Textbook Reading
- This lab appears on page **387** of the Engineering Our Digital Future textbook.
- Prerequisite textbook reading before completing this lab: pp. **384-387.**

Engineering Designs and Resources
Continuation of Lab 7.1 and 7.2

Worksheet Description
Also, unless your alone in the room with your lab partner there are probably other Air Modems which are interfering with your signal. You've probably already agreed to stop interfering by either taking turns or by adjusting the Starting Frequencies to use different frequencies for your tones. Let's examine the effectiveness of using different frequency tables simultaneously.

Laboratory Procedures

Steps	Instructions
Step 6:	Team up with another Air Modem pair near your station. Set the tone separation for both teams to 100 (remember this value needs to be set at both the transmitter and the receiver). Set team #1's Starting Frequency to 300 Hz and Team #2's Starting Frequency to 350 Hz.
	Q15: Write down the frequencies used by both teams to send the letters A B and C below: Letter Team #1 Frequency Team #2 Frequency A B C
Step 7:	Simultaneously transmit messages.
	Q16: Do they seem to be interfering?

	Reduce the Starting Frequency of Team #2 by increments of 5 Hz at a time (remember to hit enter after typing the value).
	Q17: At what point do they begin to interfere?

Overview Questions

A: How many tones were required for a single Simple Air Modem?

B: How many filters were required?

C: How easy is the character decision process?

D: At a tone spacing of 75 Hz, how many of these systems can operate simultaneously in the audio range of 300 Hz to 20 KHz.

Summary

The simple air modem uses a single frequency for each character. It is easy to understand, but perhaps not the most efficient use of transmission bandwidth. We noted the large number of frequencies required to use this method. This limits the number of systems that can operate simultaneously in parallel. We also noted that audio noise plagues systems that use the audio band to communicate. In the following labs we will investigate more efficient approaches.

7.4 Audio Communications Using Several Simultaneous Tones

Lab Objectives

In the previous lab we built a simple Air Modem that used a single frequency for each character. It is easy to understand, but perhaps not the most efficient way. We know that it only takes Log_2N "bits" to represent N things, so why to we have to use N frequencies for N characters? We don't. In this lab we'll look at ways to reduce the number of frequencies used. Figure 7.7 is an air modem approach we will term Binary Parallel. It uses presence or absence of each of the next larger integer of $Log_2(N+1)$ frequencies to represent the binary equivalent the character. In this Air Modem several frequencies are transmitted at once, each representing a different digit position of the binary representation of the character. As Figure 7.8 shows there are fewer frequencies and therefore fewer filters than the simple air modem approach but now the decision logic is more complicated as it must decide on the presence or absence of several frequencies at once.

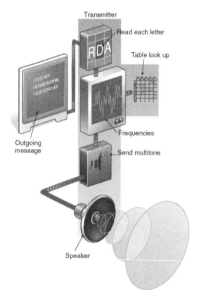

Figure 7.7 Binary Parallel Air Modem Transmitter

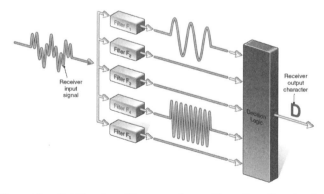

Figure 7.8 Filter Bank for Binary Parallel Receiver. There are fewer filters than the simple Air Modem but now the Decision Logic must be able to identify the presence of multiple frequencies

Engineering Designs and Resources

Worksheets used in this lab:

- **L07-04-01 Audio Communication Using Several Tones Transmit.Lst**: Transmits contents of a text file using combinations of five different tones.

- **L07-04-01 Audio Communication Using Several Tones Human Receiver.Lst:** Transmits input from the keyboard, using combinations of five different tones.

Worksheet Description

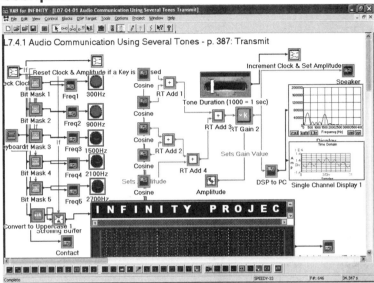

Figure 7.9 Audio Communication Using Several Tones - Transmit

Transmitter Worksheet

In this lab, you and your lab partner will construct a Binary Parallel Air Modem. One of you will run the transmitter worksheet, and the other will run the receiver worksheet.

Open Worksheet *L07-04-01 Audio Communication Using Several Tones Transmit.Lst*. You will see something similar to Figure 7.9 above.

As in the previous lab, you will see the keyboard input, the Time-domain and Frequency-Domain signal plots, the Tone Duration Slider, the textual message output, and the Calibration Curve. New to this Air Modem are the 5 parallel bit masks generating the 5 different frequencies you will use. An LED indicates the presence or absence of each of these frequencies in the current signal. First let's use the transmitter worksheet by itself.

Laboratory Procedures

Steps	Instructions
Step 1:	Start your worksheet.
Step 2:	Slowly type your name.
	Q1: How many LEDs are "lit" for each letter of your name?

Steps	Instructions (Continued)
	Q2: How many peaks are there in the spectrum of each letter?
	Q3: Do the peaks correspond to the values of the LED that are "lit"?
	Q4: Why are 5 frequencies needed?

Human Receiver Worksheet

As the receiver for the Binary Parallel Air Modem is sufficiently complex we will use a diagnostic tool for you to see the different frequencies to understand how this version of the air Modem works.

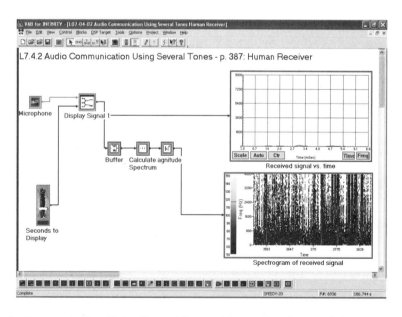

Figure 7.10 Audio Communication Using Several Tones - Human Receiver worksheet

Open the worksheet *L07-04-02 Audio Communication Using Several Tones Human Receiver.Lst*. You will see something similar to Figure 7.10 above.

You will see that it contains a microphone, a time-based signal display and a Spectrogram of the input. Recall that a vertical slice of a spectrogram is the spectrum of the signal at a given instant in time. A single frequency will be evident as a vertical slice with one colored "peak". Every frequency evident in the signal will correspond to one colored "peak" at that time. Now we are ready to use the two worksheets together.

Laboratory Procedures

Steps	Instructions
Step 3:	Type your name into the transmitter worksheet again. When you have finished typing, halt the receiver worksheet.
	Q1: Can you distinguish the peaks corresponding to the different frequencies of the letters of the name?
Step 4:	Try typing a short sentence into the transmitter.
	Q2: Can you determine what was transmitted? (Hint: you'll need to know the values of the frequencies as well as the ASCII table from the book. The transmitter sends 5 least significant bits of the ASCII code.)
	Q3: What is your method to determine what letters were sent?

Overview Questions

A: Do you think it is easier or harder to determine which character was sent in this parallel air modem compared to the simple one frequency per letter air modem?

B: How many frequencies are used by your link?

C: At a tone spacing of 75 Hz, how many of these parallel binary systems should be able to be operated simultaneously in parallel in the audio range of 300 Hz to 20 KHz.

Summary

In previous labs we examined a simple air modem in which each character is represented by a unique frequency. In this lab we saw that a binary representation of each character would allow a reduction in the number of frequencies required. In the lab that follows we will take this reduction in numbers of frequencies one step further.

7.5 Audio Communications Of Messages Using Serial Binary Transmission

Lab Objectives

We greatly reduced the number of frequencies used in each transmission system. Can we do better? Binary numbers only use 2 symbols to represent a multitude of things. We need only introduce the concept of separating the digits in our transmission system to take advantage of this. In the binary parallel air modem all digits were transmitted at the same time. The presence or absence of a frequency indicated the binary digit corresponding to that frequency being a one or a zero. Instead, let's pick one frequency to mean "one" and another frequency to mean "zero". Now we will play the binary digits sequentially "in time". The transmission of a character will take longer (multiply by the number of binary digits) but will only use two frequencies. The concept behind this is indicated in Figure 7.11 representing the filter bank at the receiver of such a system. Only two frequencies are used and a time based signal of several "bits" represents the character.

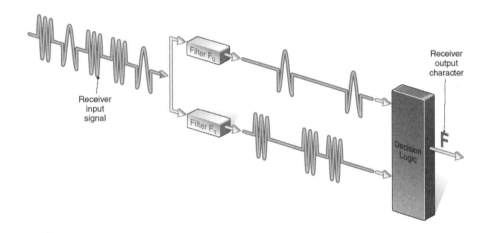

Figure 7.11 Filter bank for Binary Serial Air Modem

Textbook Reading

- This lab appears on page **393** of the *Engineering Our Digital Future* textbook.
- Prerequisite textbook reading before completing this lab: pp. **390-391**.

Engineering Designs and Resources

Worksheet used in this lab:

- **L07-05-01 Audio Communication of Messages Using Serial Binary - Transmit:** Converts each character of a textual input into a pulse train of 2 frequencies.

Worksheet Description

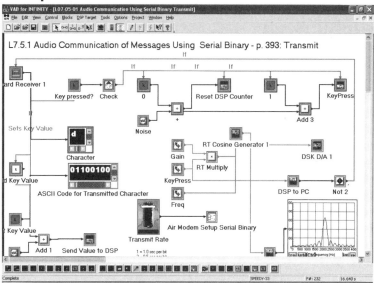

Figure 7.12 Audio Communication of Messages Using Serial Binary Transmit

Open the worksheet *L07-05-01 Audio Communication of Messages Using Serial Binary Transmit.Lst*. Your screen will show something similar to Figure 7.12.

Again characters are represented as binary digits, but rather than sending all the digits at once, they are sequentially transmitted. The keyboard takes in a character and displays its textual version in the "Character" display. A Binary version of the character is displayed in the "ASCII Code for Transmitted Character" display. The "transmit rate" slider determines the rate at which the characters and bits are transmitted. A frequency-domain plot of the transmitted signal is in the lower right-hand corner of the worksheet. The remainder of the worksheet generates the multiple bursts of two frequencies corresponding to the binary representation of the signal.

Laboratory Procedures

Steps	Instructions
Step 1:	Start your worksheet
Step 2:	Type a letter.
	Q1: What did you hear?
	Q2: How many frequency bursts were there? (Repeat the letter if needed.)

Steps	Instructions (Continued)
	Q3: Which bursts were of higher pitch, which were of lower pitch?
	Q4: How does this correspond to the "ASCII Code for Transmitted Character" Display?
Step 3:	Examine the output of the signal with the Human Receiver. Type short messages and halt the worksheet.
	Q5: How many peaks does the Human Receiver worksheet show at any instant in time?
	Q6: How many frequencies does the receiver detect across the entire message?

Overview Questions

A: Will it take longer or shorter to transmit a message in this serial air modem than the parallel binary air modem? (Assume the length of each burst is the same).

B: Do you think it is easier or harder to determine which frequency was sent in the parallel air modem compared to the simple one frequency per letter air modem?

C: How many frequencies are used by your binary serial air modem?

D: At a tone spacing of 75 Hz, how many of these binary serial systems can operate in parallel in the audio range of 300 Hz to 20 KHz.

E: Complete the table below for the three air modems we have created: The simple Air Modem, Binary Parallel Air Modem, and Binary Serial Air Modem?

Air Modem	# Frequencies	Difficulty of Decision Logic	# of parallel systems that fit in the audio band 300 Hz - 20KHz
Simple			
Binary Parallel			
Binary Serial			

Summary

Through the incremental advances made in our Air Modems we have seen that ingenuity can lead to more efficient use of available bandwidth and less expensive designs.

7.6 Touchtone Telephone

Lab Objectives

Almost every day you use multiple frequencies to transmit information. Whenever you pick up a telephone and press its keys, you are sending 2 specific frequencies down the telephone line to the phone company indicating which digit you want to "dial". Figure 7.13 shows the keypad and the frequency layout. Each row and each column have a unique frequency. When you press a key, tones corresponding to each of the two different frequencies are added together to make the resultant signal. The row and column frequencies are carefully chosen to make it clear which key you pressed. In this lab you will experiment with worksheet which creates the same tones the telephone company uses.

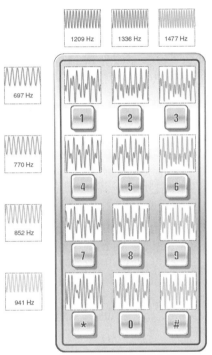

Figure 7.13 The Touchtone telephone uses two tones, one for the row and one for the column, simultaneously transmitted down the phone line to indicate the number you press.

Textbook Reading

- This lab appears on page **395** of the *Engineering Our Digital Future* textbook.
- Prerequisite textbook reading before completing this lab: pp. **393-395.**

Engineering Designs and Resources

Worksheets used in this lab:

- **L07-06-01 Touchtone Telephone Transmit.Lst**: Simulates a Touchtone keypad.
- **L07-06-02 Touchtone Telephone Human Receiver.Lst**: Uses a spectrogram to let the user determine the transmitted data.

7.6.1   Touchtone Telephone Transmit

Worksheet Description

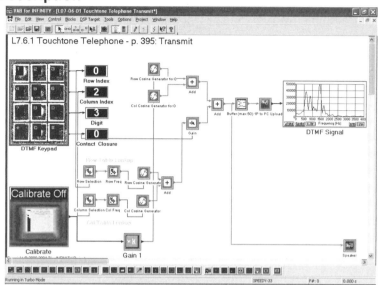

Figure 7.14 Touchtone Telephone Transmit

Open worksheet *L07-06-01 Touchtone Telephone Transmit.Lst*. You will see something similar to Figure 7.14 on your screen.

The DTMF keypad in the upper right is just like your phone keypad. The Calibrate button allows the signal for the number zero to be held for an extended period without a person holding a button on the keypad down. The numeric displays indicate the row, column, and number that was pressed as well as indicating if it is currently being pressed. The row and column indices are used to look up the appropriate frequencies in a table in the computer's memory. The two sinusoids are then generated, added together and emitted from the speaker. The spectrum of the resultant signal is displayed in the Plot.

Laboratory Procedure

Steps	Instructions
Step 1:	Start the worksheet.
Step 2:	Using the mouse to click on the DTMF keypad and dial your phone number.
	Q1: Did it sound right?
Step 3:	Have your lab partner use the Human Receiver worksheet from the previous lab to examine the numbers as they are dialed.

Steps	Instructions (Continued)
	Q2: Can you identify the row frequencies?
	Q3: What about the column frequencies?
Step 4:	If you have a land-line phone or an answering machine that will respond to touch tone phones, you can try making phone calls it with this worksheet. Hold the receiver up to the speakers and dial. Note: This won't work for cell phones; they never give you a dial tone.

Overview Questions

A: How many frequencies are used to dial each number on a touchtone telephone?

B: How many different sinusoid frequencies are needed to dial all of the numbers on a touch-tone telephone?

Summary

The signaling system the telephone company uses is a straightforward two-tone system, one tone for each row and one for each column. This can be extended to larger character sets as seen in the next lab.

7.7 Audio Communication Using Two Tones

Lab Objectives

The dual tone telephone system approach we used in the previous lab can be extended to send text messages. Figure 7.15 shows one approach to this encoding. In this lab we will explore a worksheet that implements this dual tone system.

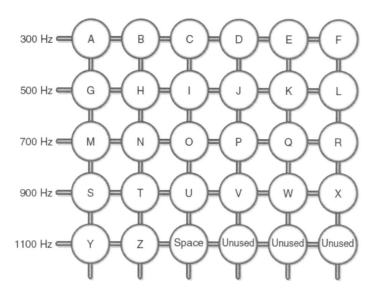

Figure 7.15 Dual Frequencies for the alphabet

Textbook Reading

*   This lab appears on page **395** of the *Engineering Our Digital Future* textbook.
*   Prerequisite textbook reading before completing this lab: pp. **393-395.**

Engineering Designs and Resources

Worksheet used in this lab:

*   **L07-07-01 Audio Communication Using Two Tones Transmit.Lst**: Transmits input from the keyboard, using combinations of two different tones.

Worksheet Description

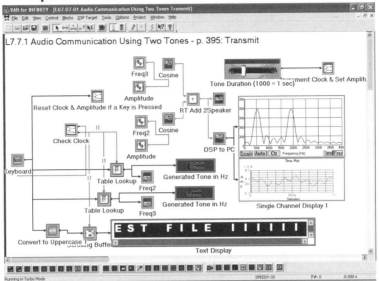

Figure 7.16 Audio Communication Using Two Tones Transmit

Open the worksheet *L07-07-01 Audio Communication Using Two Tones Transmit.Lst.* You will see something similar to Figure 7.16.

This worksheet extends the touchtone telephone worksheet we just used to transmit text messages generated from the keyboard. The tone duration can be set by the slider in the upper right-hand corner of the worksheet. The time-domain and frequency-domain plots of the signal are displayed along with the two current frequencies being transmitted.

Laboratory Procedures

Steps	Instructions
Step 1:	Start the worksheet.
Step 2:	Type a short message.
	Q1: How many peaks do you see in the spectrum of the transmitted signal?
	Q2: Is it easy for you to discern the characters that are being sent?

Steps	Instructions (Continued)
	Q3: For this approach to work, the letters must be laid out on some form of rectangular grid. If we must have spots for 26 letters and the space bar, what aspect ratios of rectangles with integer length sides have enough spaces for the characters (i.e., a 1x27 rectangle would, as would the 5x6 rectangle in Figure 7.15)?
	Q4: For each of the rectangles you identify, how many distinct frequencies would be required (i.e., number of rows + number of columns)?
	Q5: Do you think it is a coincidence that the layout of figure 7.15 is approximately square?

Overview Questions

A: If you pick 1x27, how does it compare with the "one-tone-per-character" method?

B: Could you reduce the number of tones required through the use of a 3-D scheme (e.g., 2 x 2 x 8 or 2 x 3 x 3)? How many tones would be transmitted at once in this scheme?

Summary

This worksheet extends the touchtone telephone worksheet we just used to transmit text messages generated from the keyboard. We examined the design process of picking the aspect ratio of the encoding layout.

7.8   How a Facsimile Works

Lab Objectives

In this lab we will examine the basics behind the operation of a FAX machine. The input is considered a black and white image, i.e. each pixel contains only one bit of information. The input image is raster scanned and if the pixel is black, then a tone is emitted. The image is reassembled on the receiver side mimicking the raster scan process and turning a pixel black if a tone it received, and making it white if no tone is received. We can use 1 bit images, or user entered data for the input. Also we can transmit the data over a network to another workstation.

Textbook Reading

*   This lab appears on page **401** of the *Engineering Our Digital Future* textbook.
*   Prerequisite textbook reading before completing this lab: pp. **397-401.**

Engineering Designs and Resources

Worksheets used in this lab:

*   **L07-08-01 How a facsimile works Image.Lst**: Allows user to simulate the faxing of a bitmap on their hardware.
*   **L07-08-02 How a facsimile works User Image.Lst**: Allows user to simulate the faxing of an image they can edit.

7.8.1   How a facsimile works - Image

Worksheet Description

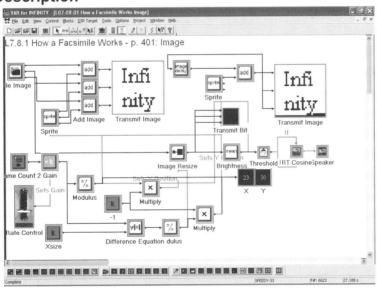

Figure 7.17 How a facsimile works - Image

Open worksheet *L07-08-01 How a Facsimile Works Image.Lst* and you will see something similar to Figure 7.17 above except the displays will be blank.

In the upper left-hand corner is a File Image read block which gets the input image from your computer's hard drive. It must be a 32x32 image file containing only black (0) and white (255) pixels. If you wish to use a different size image you can modify the PC variable Xsize to have a different value than 32.

The blocks in the lower half of the worksheet calculate the position of the raster scanned cursor in the original image. The frame counter block counts how many times all the blocks in the sheet have been executed. The Difference Equation block, Mod blocks, and Multiply blocks that follow it convert this into an X and Y position of the cursor on the original image. The image resize block crops the original image to a 1x1 pixel image located at the cursor which is displayed in the transmit bit display (white or black). That average brightness of that 1x1 image (i.e., its value) is thresholded to determine whether the cursor is currently on a white or black pixel. If the pixel is black, then a tone is emitted from the speaker, otherwise the speaker is silent. One the received image display the cursor follows the raster scanned image and replaces the pixel value with current pixel value. This is similar to the operation of a Fax machine.

The main difference being that a fax machine takes advantage of the fact that relatively few pixels are black and performs compression on the image before it is transmitted. Both this worksheet and an actual fax machine use signals in the audio band to transmit the pixel values of the image.

Laboratory Procedures

Steps	Instructions
Step 1:	Start the worksheet.
	Q1: Describe the pattern the red cursor is taking on the original image.
	Q2: What do you hear from the speaker?
	Q3: What is the color of the pixel when you hear a tone?
Step 2:	Halt the worksheet
Step 3:	Adjust the Rate Control slider and Reset and Compile the worksheet
Step 4:	Run the worksheet.

Steps	Instructions (Continued)
	Q4: How did it change?
	Q5: Estimate the percentage of time that the speaker is silent for the default image.

7.8.2   How a facsimile works - User Image

Worksheet Description

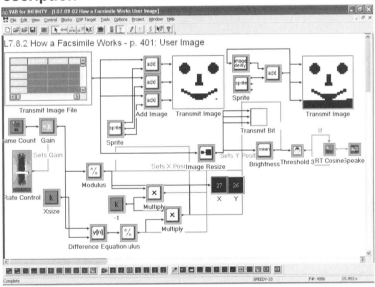

Figure 7.18 How a facsimile works - User Image

Open the worksheet *L07-08-02 How a Facsimile Works User Image.Lst.* You will see something similar to Figure 7.18 above, only the displays will be blank.

Now we will allow you to edit your own image. This worksheet will allow you to edit the input image which the FAX machine sends. Recall that a digital image is simply a matrix of numbers. We can edit the matrix of numbers in the upper left to create any black and white input image. You should only enter the numbers 255 and 0 in the matrix as the fax machine can only handle black and white images.

Laboratory Procedures

Steps	Instructions
Step 1:	Start the worksheet.
Step 2:	Edit the matrix of numbers to create your own image.
	Q1: Can you make the fax machine print your name in the received image?
Step 3:	Try other simple images.
	Q2: Can you deduce any characteristics of the image by how it sounds?

Summary

In this lab you have examined the basics behind the operation of a FAX machine.

Networks From the Telegraph to the Internet

Chapter 8 gives an overview of computer networks and the Internet from both modern and historical perspectives. In these labs students will construct simple networks and use them to transmit messages. Students can take turns being users of the network or routers within the network.

Infinity Labs

Introduction

The laboratories of Chapter 8 have you incrementally build up a multi-user network to understand the constructs and roles of the computers that comprise the internet. A simple multi-user text chat session is the application for which first a fully meshed network, followed by a star network, then hierarchical topologies are created. You will see the work that must be done by different parts of the network as well as what sort of services are required.

The basic input blocks of Chapter 8 are keyboard input, as the application is a multi-user text chat. The output is a set of text message displays. Much of the educational goals of this chapter is achieved during the thinking and planning for the network rather than its operation. A functional network is simply an indication of correct planning.

8.1   Multiple-User Network with Meshed Connections

Lab Objectives
The digital world is a connected one. Computer networks, such as the Internet, link devices like airplanes and vending machines so that they can share information with each other. How do these networks work? In this and following labs, we'll learn about network topologies – the physical or virtual structure of the network connections – to help us understand how computer networks are designed. The first network we will learn about is a multi-user network with meshed connections, laid out in the form of a four-user text chat system.

Textbook Reading
- This lab appears on page **413** of the *Engineering Our Digital Future* textbook.
- Prerequisite textbook reading before completing this lab: pp. **411-412.**

Engineering Designs and Resources
Worksheet used in this lab:

- **L08-01-01 Multiple-User Networks with Meshed Connections.Lst**: A simulated meshed topology network for multiple users.

Worksheet Description
In this lab we will build a series of network topologies to help you understand how networks function and what the role and procedures for routers are. Figure 8.1 below is a picture of the first network we will construct. It contains only 4 users, each with a direct connection to the 3 other users. As we know from the text, in practice users are usually connected through routers. This worksheet will let you get the basics down so you can construct more realistic networks with routers later.

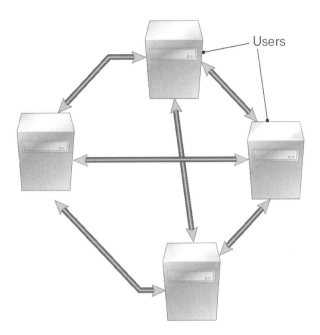

Figure 8.1 Fully Meshed Network containing 4 users

To build this network you will need 4 users (yourself and three lab partners). Each of you will use the same worksheet: *L08-01-01 MUN with Meshed Connections.Lst* shown in Figure 8.2.

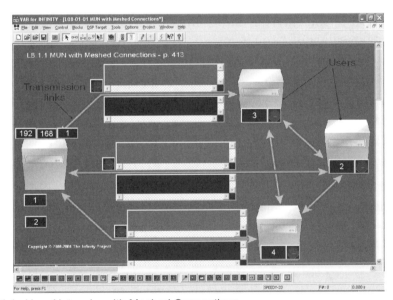

Figure 8.2 Multiple-User Networks with Meshed Connections

In this worksheet, the person running the sheet is represented by the user computer on the far left. The three users on the right represent the computers of your lab partners. Along each link between you and one of your lab partners are two text boxes: one that contains text you sent to that lab partner, and one that contains text received from that lab partner. The status lights next to the text boxes indicate which lab partner you are currently sending text to, and the status lights next to the users on the right indicate which users are currently transmitting text to you. The various numeric entry boxes around the sheet hold the network address information which we will now configure below:

Network Configuration

Before we begin to use this worksheet we need to configure the worksheets by entering all the addresses of the computers involved in your network.

1. Your instructor may be able to provide you the IP address of your computer and your partners. It will be in the form of 4 numbers (each less than 256). For example a valid address might be 192.168.1.1. You can obtain the address yourself in Windows by clicking on the **Windows Start** button, selecting **Run** and typing "command" and hitting **OK**. This will open a command prompt window. In this window you can type the command "ipconfig" and it will report various information about your network configuration including your IP addresses.

   You'll need to know your IP address and the address of your 3 lab partners. For this example let's assume your IP address is 192.168.1.1 and your partners are 192.168.1.2; 192.168.1.3 and 192.168.1.4 respectively. This lab worksheet assumes you are on the same local network and therefore the first three numbers of each of your IP addresses are the same.

2. Open the worksheet, but **don't run it yet** or it will hang up trying to establish connections to non-existent IP addresses.

3. Above your user on the left hand side of the worksheet you will see 3 numeric entry boxes. Enter the first three numbers of your IP address in these boxes (e.g., 192, 168, 1). You will need to hit **enter** after each number you enter for it to be accepted by the worksheet.

4. Below your user on the left hand side of the worksheet you will see 2 numeric entry boxes, one above the other. In the top box enter the last digit of your IP address, and don't forget to hit enter.

5. Next, enter the last digit of each of your lab partner's IP addresses in the numeric entries associated with the 3 users on the right hand side of the worksheet. Again, don't forget to hit **enter** after each one.

6. The final numeric entry box (bottom left corner of the worksheet) will be used to indicate the user to which you want to send text. Whichever user matches that number will receive the messages you type.

When you and your lab partners have completed this, you are ready to start the lab.

Laboratory Procedures

Steps	Instructions
Step 1:	Make sure you and your lab partners have all configured your worksheets following the instructions above.
	Enter one of you lab partner's address (fourth number) in the numeric entry in the bottom left-hand corner of the worksheet.
Step 2:	When each of you are ready, start your worksheets.
Step 3:	Begin typing a message.

Steps	Instructions (Continued)
	Q1: Did it appear next to the correct link in your worksheet (the one connected to the user you specified)?
	Q2: Did he or she receive the message?
Step 4:	Choose a different lab partner to receive your messages and enter his/her fourth digit address in the lower left-hand numeric entry box.
	Q3: Were you able to successfully change which user received your message?
	Q4: What would happen if you had an incorrect address for your lab partner?
Step 5:	Continue chatting with your friends over your fully meshed network for a bit, then consider the overview questions below:

Overview Questions

A: How many virtual links were there in your 4 user network?

B: How many feet of wire would have been required to set up this fully meshed network? You can assume that all wires are the same length as an average distance between your computers.

C: Repeat those calculations for the total number of people in your class, and your school.

D: Why do you think the internet isn't configured this way?

Summary

A full mesh network is great for small network topologies, but it is very expensive for large ones. Suppose you wanted to connect a city of 1 million people together. How many "wires" would you need in a full mesh network? Is such a network practical? In the labs that follow, we'll look at some more realistic network topologies that have a lot lower wiring cost.

8.2 Multiple User Network Using a Single Router

Lab Objectives

In the last lab, you simulated a full mesh network in the form of a text chat with up to 3 other friends. In this lab, you'll simulate a chat system with 2 friends with a slight twist: one user will relay messages for everybody else. This type of network, called a star configuration network, has its advantages and disadvantages, as you will see — and you might even get to be the router!

Textbook Reading

*   This lab appears on page **415** of the *Engineering Our Digital Future* textbook.
*   Prerequisite textbook reading before completing this lab: pp. **413-414.**

Engineering Designs and Resources

Worksheets used in this lab:

*   **L08-02-01 Multiple-User Networks Using a Single Router User.Lst:** Allows you to be a user in a star configuration network between computers in the lab.
*   **L08-02-02 Multiple-User Networks Using a Single Router.Lst:** Allows you to be the router in a star configuration network between computers in the lab.

Worksheet Description

In the last lab you built a fully meshed network of 4 users, and you did some calculations to figure out why the internet isn't configured exactly this way. In this lab, we will use a router to connect 3 users. The configuration of our new network is depicted in Figure 8.3 below.

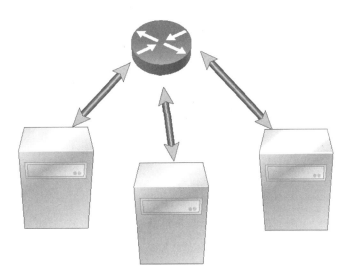

Figure 8.3 Three users connected through a single router.

To build this network you will need 4 people (yourself and three lab partners). You should decide who will be the router and who will be the three users. You can should take turns being a router so you each get to see how much work is involved.

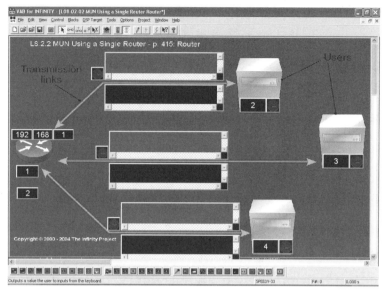

Figure 8.4 Multiple-User Networks Using a Single Router - Router

In this lab the router will use the worksheet *L08-02-02 MUN Using a Single Router - Router.Lst*. As in Lab 8.1, you (the router) are on the left-hand side of the screen and your users (lab partners) are on the right-hand side. You should follow the steps of Network Configuration in Lab 8.1 to enter the appropriate network addresses in this sheet.

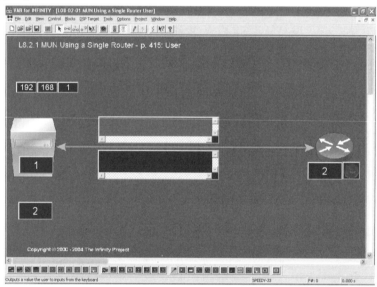

Figure 8.5 Multiple-User Networks Using a Single Router - User.Lst

The users in this lab will use the worksheet *L08-02-01 MUN Using a Single Router User.Lst*. This worksheet is a lot simpler than that in Lab 8.1, because you are only allowed to communicate with your router now. You should also follow the steps of Network Configuration in Lab 8.1 to enter the addresses of yourself and your router.

Name Server Table Preparation

Since in this lab users will not communicate directly with one another and the router will only be able to route messages based on IP addresses, users will need a method to convert the name of the person they want to communicate with an IP address. This function is called Name Service Lookup and we will perform it by constructing a table to convert the names of the users to IP addresses. Fill out the Name Services Table below. It should contain the 3 users (including your-self) and their IP addresses. Since you all share the first 3 digits of you IP addresses you need only enter the 4th digit In the table below enter the fourth digit address for yourself and the 3 users in the network.

Name Service Table

User Name	Fourth Digit of IP Address

The next thing you'll need to agree on is a way of knowing who you are intending to talk to and who you are receiving messages from. All the users' interactions will be with the router, but this is sort of like saying all mail comes from the mail carrier. Collectively you should agree to address all messages and indicate a return address. One method that might work is to begin every mes-sage with the IP address of the person you want to talk to, then a colon, then your IP address. For instance, suppose Tom's 4th digit address is 10 and Sally's is 12. Tom wants to send Sally a message through the router he can send the message "12:10: What's up?". It's a good idea to put the To: person first as it will be the first thing the router sees.

Laboratory Procedures - Router

Steps	Instructions
Step 1:	Make sure you and your lab partners have all configured your worksheets following the instructions above. Enter one of you lab partners address (fourth number) in the numeric entry in the bottom left-hand corner of the worksheet.
Step 2:	When each of you are ready, start your worksheets.
Step 3:	As the router, you have nothing to do until someone sends you a message. Don't worry, you'll be plenty busy once that starts.
Step 4:	When someone sends you a message it will appear in the text box below the link to them on your worksheet. If the message has an obvious destination based on your agreed upon protocol, enter their fourth digit address in the lower left-hand numeric entry box, and retype the message to the intended recipient. If the message does not have an obvious recipient (i.e. They didn't follow the agreed upon protocol) respond to the sender of the message with an error.
Step 5:	Do your best to keep up with the users. When your fingers get tired of typing sug-gest letting someone else be the router for a while.

Laboratory Procedures - User

Steps	Instructions
Step 1:	Make sure you and your lab partners have all configured your worksheets following the instructions above. Enter your router's address in the lower left-hand numeric entry box. This is the only person you can communicate directly with.
Step 2:	When each of you are ready, start your worksheets.

Steps	Instructions (Continued)
Step 3:	Type a message to another user. Don't forget to follow the protocol you agreed upon as a group.
Step 4:	Choose your recipients at will among the users in your group and carry on a text chat with them. Don't overload your router or data will get lost. Send the messages in small pieces.

Overview Questions

A: How many links were required to set up this network?

B: How much wire would this network require?

C: How fast did this network work compared to a fully meshed network?

D: Which of you were the busiest: the users or the router?

E: Given your answers to the prior 4 questions, why do you think the internet uses routers?

F: What is a star configuration network?

G: What is the main advantage of the star configuration network?

H: What is the main disadvantage of the star configuration network?

Summary

A full mesh network is great for small network topologies, but it is impractical for large ones. In this star configuration fewer links were required but at the expense of requiring a router to do a lot of work. In the labs that follow, we'll look at some more realistic network topologies.

8.3 Multiple User Network Using Several Routers

Lab Objectives

The first two labs of this chapter introduced you to full meshed and star configuration networks, respectively. This lab shows you a network topology that more closely resembles that of the Internet: A multiple-router network. Sending messages requires some know-how as to who lives where on the network-and that's where routing tables come in.

Textbook Reading

*   This lab appears on page **416** of the *Engineering Our Digital Future* textbook.
*   Prerequisite textbook reading before completing this lab: pp. **415-416.**

Engineering Designs and Resources

Worksheets used in this lab:

*   **L08-03-01 MUN Using Several Routers User.Lst:** Allows you to be a message sender and receiver in a multi-relay network between computers in the lab.
*   **L08-03-02 MUN Using Several Routers Outside Router.Lst:** Allows you to act as a relay in a multi-relay network between computers in the lab.
*   **L08-03-02 MUN Using Several Routers Central Router.Lst:** Allows you to act as a relay in a multi-relay network between computers in the lab.

Worksheet Description

In the last lab you built a network of 3 users connected through a router. This time you will see how to scale that network to larger sizes. When the network scales to larger sizes, often there are routers which only communicate with other routers. They make up the backbone of the network. The configuration of our new network is depicted in Figure 8.6 below.

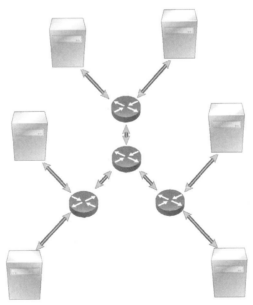

Figure 8.6 Six users connected through a hierarchy of 4 routers.

To build this network you will need 10 people (yourself and nine lab partners). You should decide who will be the central router (the router that only communicates to other routers), each of the three outside routers (which communicate to the central router and 2 users) and who will be the six users. You should take turns in each role so you each get to see how much work is involved.

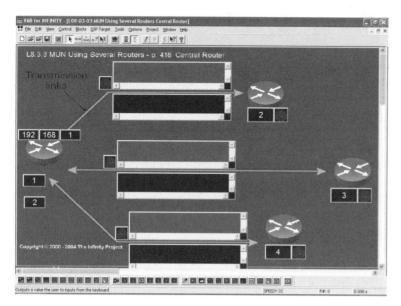

Figure 8.7 Multiple-User Networks Using Several Routers Central Router

In this lab, the central router will use the worksheet *L08-03-03 MUN Using Several Routers Central Router.Lst*. As in Lab 8.1 you (the router) are on the left-hand side of the screen and the routers you communicate with (lab partners) are on the right-hand side. You should follow the steps of Network Configuration in Lab 8.1 to enter the appropriate network addresses in this sheet.

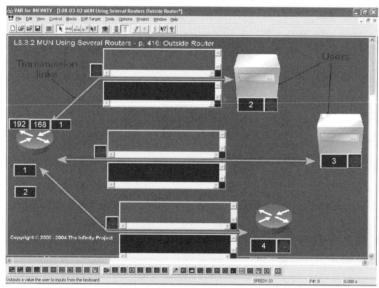

Figure 8.8 Multiple-User Networks Using Several Routers Outside Router

In this lab the outside router will use the worksheet *L08-03-02 MUN Using Several Routers Outside Router.Lst*. As in Lab 8.1 you (the router) are on the left-hand side of the screen and the router and users you communicate with (lab partners) are on the right-hand side. You should follow the steps of Network Configuration in Lab 8.1 to enter the appropriate network addresses in this sheet.

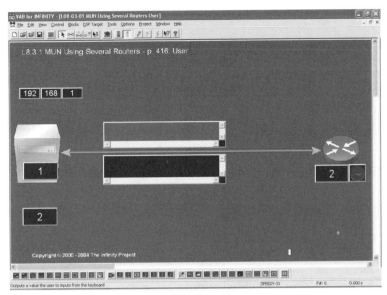

Figure 8.9 Multiple-User Networks Using Several Routers User

The users in this lab will use the worksheet *L08-03-01 MUN Using Several Routers User.Lst*. You are only allowed to communicate with your router, no direct communication with other users. You should also follow the steps of Network Configuration in Lab 8.1 to enter the addresses of yourself and your router.

Router Table Preparation

As in the previous lab, you and your lab partners need to agree to use some protocols and provide the routers with the information they need. The first thing you'll need to agree on is a way of knowing who you are intending to talk to and who you are receiving messages from. Collectively you should agree to address all messages and indicate a return address. The router needs a table to convert the destination name into the address of the next hop. In this lab, a router is not necessarily connected directly to the user who is the ultimate destination of the message. A routing table will be used with a row for each user and a Next Hop address that tells the router where to send the message. As there are 6 users there are 6 rows in the routing table below. The group should determine the routing table for each router in the group and fill in the next hop for each destination. This routing table will be different for each router in the network. If you are a router, fill in the table below for the specific router you will be. Remember to modify the table if you change to a different router later.

Router Table

Message Destination	Next Hop Fourth Digit Address

You will still need a Name Service Table just like in Lab 8.2 Fill out the Name Service table below for the 6 users.

Name Service Table

Message Destination	Next Hop Fourth Digit Address

The users will use only the name service table (to determine the address of the destination of their message), whereas the routers will only use only their specific router table (to route the messages they receive).

Laboratory Procedures - Router

Steps	Instructions
Step 1:	Make sure you and your lab partners have all configured your worksheets following the instructions above. Enter the address (fourth number) of a node you communicate directly with in the numeric entry in the bottom left-hand corner of the worksheet.
Step 2:	When each of you are ready, start your worksheets.
Step 3:	As the router, you have nothing to do until someone sends you a message.
Step 4:	When someone sends you a message, it will appear in the text box below the link to them on your worksheet. If the message has an obvious destination based on your agreed-upon protocol, look up the recipient in your router table (above), enter the fourth digit address for the next hop in the lower left-hand numeric entry box, and retype the message to the intended recipient. If the message does not have an obvious recipient (i.e. they didn't follow the agreed-upon protocol) respond to the sender of the message with an error.
Step 5:	Do your best to keep up with the users. When your fingers get tired of typing suggest letting someone else be the router for a while.

Laboratory Procedures - User

Steps	Instructions
Step 1:	Make sure you and your lab partners have all configured your worksheets following the instructions above. Enter your router's address in the lower left-hand numeric entry box. This is the only person you can communicate directly with.
Step 2:	When each of you are ready, start your worksheets.
Step 3:	Type a message to another user. Don't forget to follow the protocol you agreed upon as a group.
Step 4:	Choose your recipients at will among the users in your group and carry on a text chat with them. Don't overload your router or data will get lost. Send the messages in small pieces.

Overview Questions

A: How many links were required to set up this network?

B: How much wire would this network require?

C: How fast did this network work compared to a fully meshed network?

D: Which of you were the busiest: the users, the outside router, or the central router?

E: Based on your answer to Q4, which router needs to be of the highest performance for the network to run smoothly?

F: What is a multi-relay network?

G: For each individual user in the network you simulated, how many connections need to be "negotiated" or set up for each user? For each router?

H: What is the main advantage of this network configuration over the star configuration?

I: Give a disadvantage of this particular network.

Summary

Now, you have simulated a network similar to that of the Internet, with several routers in between users. But there's one additional wrinkle we've yet to explore: the possibility of more than one path between users in the network.

8.4 Multiple-User Network With Choice of Transmission Path

Lab Objectives

In this lab we will see that as we scale networks like the one we built in L8.3 there are often multiple paths a message may take. We will design router tables to account for this fact. The configuration of our new network is depicted in Figure 8.10 below.

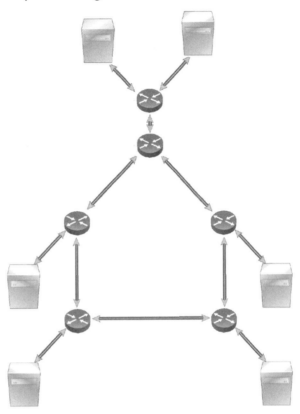

Figure 8.10 Six users connected through a hierarchy of 6 routers

Textbook Reading

*   This lab appears on page **424** of the *Engineering Our Digital Future* textbook.
*   Prerequisite textbook reading before completing this lab: pp. **419-423.**

Engineering Designs and Resources

Worksheets used in this lab:

*   **L08-04-01 MUN with Choice of Transmission Path User.Lst:** Allows you to be a message sender and receiver in a multi-relay network between computers in the lab.
*   **L08-04-02 MUN with Choice of Transmission Path Outside Router.Lst:** Allows you to act as a relay in a multi-relay network between computers in the lab.
*   **L08-04-03 MUN with Choice of Transmission Path Central Router.Lst:**Allows you to act as a relay in a multi-relay network between computers in the lab.

To build this network you will need 12 people (yourself and 11 lab partners). You should decide who will be the central router (the router that only communicates to other routers), the topmost router (which communicates with 2 users) each of the three outside routers and who will be the six users. You should take turns in each role so you each get to see how much work is involved.

Worksheet Description

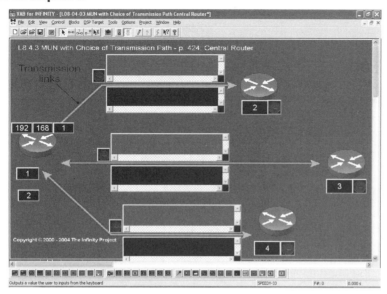

Figure 8.11 Multiple-User Network With Choice of Transmission Path Central Router

In this lab, the central router will use the worksheet L*08-04-03 MUN with Choice of Transmission Path Central Router.Lst*. As in previous labs you (the router) are on the left-hand side of the screen and the routers you communicate with (lab partners) are on the right-hand side. You should follow the steps of Network Configuration in Lab 8.1 to enter the appropriate network addresses in this sheet.

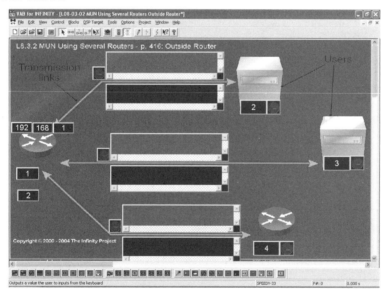

Figure 8.12 Multiple-User Network Using Several Routers Outside Router

In this lab, the topmost router will use the worksheet *L08-03-02 MUN Using Several Routers Outside Router.Lst*. As in previous labs you (the router) are on the left-hand side of the screen and the users and router you communicate with (lab partners) are on the right-hand side. You should follow the steps of Network Configuration in Lab 8.1 to enter the appropriate network addresses in this sheet.

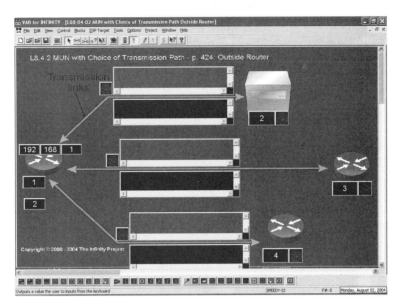

Figure 8.13 Multiple-User Network Using Several Routers Outside Router

In this lab, the outside router will use the worksheet *L08-04-02 MUN Using Several Routers Outside Router.Lst*. As in previous labs you (the router) are on the left-hand side of the screen and the router and users you communicate with (lab partners) are on the right-hand side. You should follow the steps of Network Configuration in Lab 8.1 to enter the appropriate network addresses in this sheet.

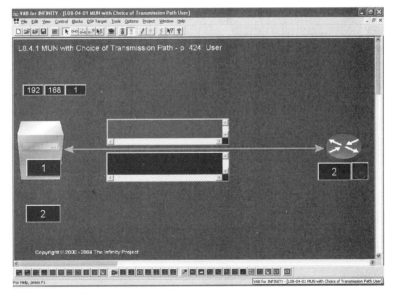

Figure 8.14 Multiple-User Network Using a Single Router User

The users in this lab will use the worksheet *L08-04-01 MUN Using a single router User.Lst* You are only allowed to communicate with your router. You should also follow the steps of Network Configuration in L8.1 to enter the addresses of yourself and your router.

Router Table Preparation

You will need to fill out the Router Table and Name Service table for this network following the procedures outlined in the previous lab. As this network has more than one path to reach any user, fill out the 3rd column of the router table with the address of the alternate path next hop.

Router Table

Message Destination	Next Hop Fourth Digit Address	Alternate Path Next Hop 4th Digit Address

Name Service Table

Name Service Table

Message Destination	Next Hop Fourth Digit Address

Laboratory Procedures- Router

Steps	Instructions
Step 1:	Make sure you and your lab partners have all configured your worksheets following the instructions above. Enter the address (fourth number) of a lab partner you communicate directly with in the numeric entry in the bottom left-hand corner of the worksheet.
Step 2:	When each of you are ready, start your worksheets.
Step 3:	As the router, you have nothing to do until someone sends you a message.
Step 4:	When someone sends you a message it will appear in the text box below the link to them on your worksheet. If the message has an obvious destination based on your agreed-upon protocol, look up the recipient in your router table (above), enter the fourth digit address for the next hop in the lower left-hand numeric entry box, and retype the message to the intended recipient. If the message does not have an obvious recipient (i.e. they didn't follow the agreed-upon protocol) respond to the sender of the message with an error.
Step 5:	Do your best to keep up with the users. When your fingers get tired of typing suggest letting someone else be the router for a while.
Step 6:	Don't forget to use the alternative path if the primary path is overworked.

Laboratory Procedures - User

Steps	Instructions
Step 1:	Make sure you and your lab partners have all configured your worksheets following the instructions above. Enter your router's address in the lower left-hand numeric entry box. This is the only person you can communicate directly with.
Step 2:	When each of you are ready, start your worksheets.
Step 3:	Type a message to another user. Don't forget to follow the protocol you agreed upon as a group.
Step 4:	Choose your recipients at will among the users in your group and carry on a text chat with them. Don't overload your router or data will get lost. Send the messages in small pieces.

Overview Questions

A: How many links were required to set up this network?

B: How much wire would this network require?

C: How fast did this network work compared to a fully meshed network?

D: Which of you were the busiest: the users, the outside router, or the central router?

E: Based on your answer to Q4, which router needs to be of the highest performance for the network to run smoothly?

F: Based on the sheets we've used can you configure a network to use your entire class?

Summary

This hierarchical network of routers and users is the basic approach taken in the design of the internet.

8.5 Internet Performance

Lab Objectives

In prior labs in chapter 8 we've seen how busy the routers of a network can get. In this lab we will examine sending large amounts of live data over the network and see the resulting delay.

Textbook Reading

- This lab appears on page **438** of the *Engineering Our Digital Future* textbook.
- Prerequisite textbook reading before completing this lab: pp. **429-437.**

Engineering Designs and Resources

Worksheets used in this lab:

- **L08-05-01 Internet Performance Audio.Lst**: This worksheet allows the user to transmit an audio file over the network to a lab partner.
- **L08-05-02 Internet Performance Audio Microphone.Lst:** This worksheet allows the user to send and receive a voice chat over the network with a lab partner.
- **L08-05-03 Internet Performance Video Transmitter.Lst:** This worksheet allows the user to send video from the web camera over the internet to a user using the Receiver worksheet.
- **L08-05-04 Internet Performance Video Receiver.Lst:** This worksheet allows a user to remotely view video sent via worksheet L08-05-03.

8.5.1 Internet Performance Audio

Worksheet Description

In this lab, we'll send audio and video over the network and notice the degraded performance due to the high demands placed by the audio and video transmission.

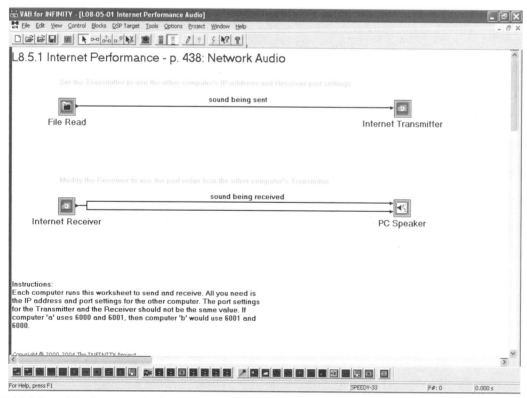

Figure 8.15 Internet Performance Audio

Laboratory Procedures - Network Audio

Steps	Instructions
Step 1:	You and your lab partner should both open the worksheet *L08-05-01 Internet Performance Audio.Lst*.
Step 2:	Double click on the internet transmitter block. Enter your lab partner's IP and select a port to transmit on. Your lab partner will have to transmit on a different port than you do.
Step 3:	Double click the internet receiver block. Enter the port number that your lab partner chose to transmit on.
Step 4:	Double click the read file block. Choose a sound file from the C:\Program Files\Hyperception\VABINF\sounds directory, Click **OK**.
Step 5:	Start your worksheet.
Step 6:	Listen to the results.
	Q1: Describe the quality of the sound file. Does it stop and start?
	Q2: Give reasons why the sound file plays the way it does.

8.5.2 Internet Performance Audio Microphone

Worksheet Description

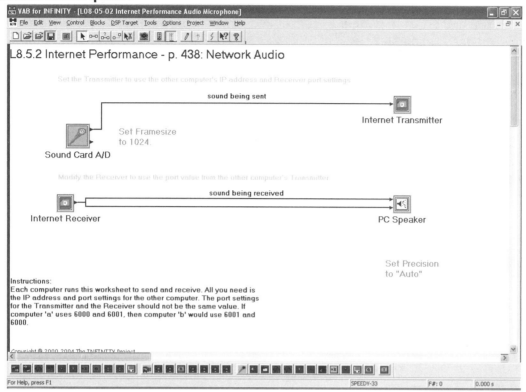

Figure 8.16 Internet Performance Audio Microphone

This worksheet is the same as the previous except a microphone is the source of the sound rather than a sound file.

Laboratory Procedures - Network Audio Microphone

Steps	Instructions
Step 1:	You and your lab partner should both open the worksheet *L08-05-02 Internet Performance Audio Microphone.Lst.*
Step 2:	Double click on the internet transmitter block. Enter your lab partner's IP and select a port to transmit on. Your lab partner will have to transmit on a different port than you do.
Step 3:	Double click the internet receiver block. Enter the port number that your lab partner chose to transmit on.
Step 4:	Start your worksheet.
Step 5:	Speak into the microphone
Step 6:	Listen to the results.

Steps	Instructions (Continued)
	Q1: Describe the quality of the sound file. Does it stop and start?
	Q2: See if you can measure the delay of the link. Begin counting into your microphone and listen for your own voice counting back to you over your lab partner's microphone. This time is the round trip delay (including the processing time of the computer).
	Q3: Give possible reasons for this delay.

8.5.3 Internet Performance Video Transmitted and 8.5.4 Internet Performance Video Receiver

Worksheet Description

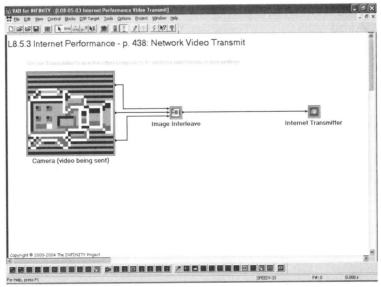

Figure 8.17 Internet Performance Video Transmit

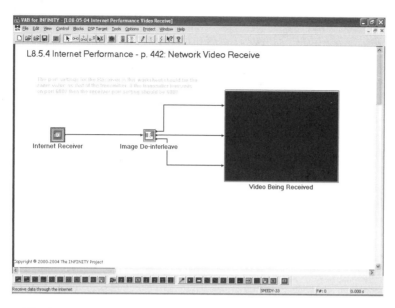

Figure 8.18 Internet Performance Video Receive

These worksheets will allow you to send video from your camera to your lab partner over the network.

Laboratory Procedures - Network Audio Video Transmit and Receive

Steps	Instructions
Step 1:	You and your lab partner should open the worksheets *L08-05-03 Internet Perfor-mance Video Transmit.Lst* and *L08-05-04 Internet Performance Video Receive.Lst*. respectively.
Step 2:	On the transmitter sheet: Double click on the internet transmitter block. Enter your lab partner's IP and select a port to transmit on.
Step 3:	On the receiver sheet: Double click the internet receiver block. Enter the port number that your lab partner chose to transmit on.
Step 4:	Start your worksheet.
Step 5:	Point the camera at something interesting.
	Q1: What do you think the "image interleave" block on the transmitter sheet is doing? (Hint: How many outputs are there from the camera versus inputs on the internet transmitter.)

Steps	Instructions (Continued)
	Q2: What does the "image deinterleave" block on the receiver sheet do?
	Q3: Try to gauge the delay of this link. Hold up your hands and count from one to 10 on your fingers. Have your lab partner respond with the same number of fingers when he/she sees your hand change. How long is the round trip delay?
	Q4: Is it different from the delay for audio?
	Q5: Why do you think that is?

Summary

In this chapter you learned how to build up networks in a hierarchical fashion and you learned about delays on network transmissions.

8.6 Exploring the Internet

Lab Objectives

In this lab we will use a standard internet tool to explore the layout of the internet.

Textbook Reading

- This lab appears on page **442** of the Engineering our Digital Future textbook.
- Prerequisite textbook reading before completing this lab: pp. **438-442**.

Engineering Designs and Resources

This laboratory will not use the VAB worksheets. It will use a utility program called TRACER-OUTE that is on PCs.

Laboratory Description

What is TRACEROUTE?

TRACEROUTE is a software program used by Internet engineers to determine how data packets flow from one point to another and to get some idea it runs to try to reach the desired destination, which is specified with either the IP address or the Uniform Resource Locator (URL). Using an interesting strategy of repeating trials, TRACEROUTE reaches deeper and deeper into the network to determine how packets will be directed to the desired destination and how long the round trip is for sending a query and getting a response. TRACEROUTE repeats this three times so that differences in the time delay measurements can be seen as well.

How do you do a TRACEROUTE?

On your PC, click on **Start**, then **Programs**, then **Accessories**, and then **Command Prompt**. You should be rewarded with old DOS prompt of the form C:\>. [If it's a different letter on your computer, don't worry about it.]

The command to perform a TRACEROUTE is TRACERT, followed by the IP address of the destination in "three dot" form, or the destination's URL. If you use the URL (which is the easiest thing to do, usually) TRACERT uses a service called Domain Name Service (DNS) to go out to a Name Server on the Internet and find the IP address associated with the URL.

Thus, if you wanted to find the path between the router that serves you, and the main router at, say, Rice University, you would type **C:\> tracert rice.edu**

```
Command Prompt                                                          _ 8 X
Microsoft Windows XP [Version 5.1.2600]
(C) Copyright 1985-2001 Microsoft Corp.

N:\>tracert rice.edu

Tracing route to rice.edu [128.42.5.4]
over a maximum of 30 hops:

  1    15 ms    15 ms    16 ms  adsl-67-118-3-254.dsl.sntc01.pacbell.net [67.118
.3.254]
  2    16 ms    14 ms    16 ms  dist1-vlan50.sntc01.pbi.net [63.203.35.65]
  3    15 ms    15 ms    15 ms  bb1-g2-0.sntc01.pbi.net [63.203.51.1]
  4    14 ms    15 ms    16 ms  bb2-p6-0.sntc01.sbcglobal.net [151.164.40.170]
  5    15 ms    15 ms    15 ms  ex1-p12-0.eqsjca.sbcglobal.net [64.161.1.38]
  6    17 ms    15 ms    15 ms  sl-st20-sj-0-0.sprintlink.net [144.223.242.81]
  7    16 ms    15 ms    15 ms  144.232.29.227
  8    15 ms    15 ms    15 ms  144.232.19.22
  9    17 ms    18 ms    15 ms  xe-0-2-0.r20.snjsca04.us.bb.verio.net [129.250.2
.72]
 10    18 ms    18 ms    18 ms  p64-0-0-0.r20.plalca01.us.bb.verio.net [129.250.
2.70]
 11    56 ms    55 ms    55 ms  p16-0-1-1.r20.dllstx09.us.bb.verio.net [129.250.
4.104]
 12    60 ms    60 ms    65 ms  p16-5-0-0.r02.hstntx01.us.bb.verio.net [129.250.
5.41]
 13    59 ms    61 ms    61 ms  ge-1-2.a04.hstntx01.us.ra.verio.net [129.250.29.
90]
 14    60 ms    60 ms    60 ms  fa-4-33.a04.hstntx01.us.ce.verio.net [128.241.2.
166]
 15    62 ms    59 ms    61 ms  walnut-v288.rice.edu [128.42.248.238]
 16    64 ms    61 ms    61 ms  moe.rice.edu [128.42.5.4]

Trace complete.
```

Figure 8.19 *tracert rice.edu* command output

The results of performing this command from a house in California are shown above. Note that the results will be different from wherever you do it, since the path through the network obviously depends on exactly where both the originator and destination are.

How do you interpret the results?

TRACEROUTE will give up trying to reach the destination if it can't find it within 30 "hops", that is, if it can't find a path through the Internet that uses 30 routers or less. In the example above it found a path from Palo Alto, California to Rice in Houston using only 16 routers.

Here's what the information in each of the columns means ---

*   Column 1: The hop number; hence line #1 provides information about the first router reached
*   Columns 2, 3, and 4: Round trip time delay, done three times
*   Column 5: Officially this is where the IP address of the router goes, but usually the network managers put more information here for their own use. As we will see, this information can give us a lot of insight about how your data will travel to the destination.

In this example, we can easily see that my e-mail to someone at Rice would travel through 15 routers to end up at one named moe.rice.edu, which has the IP address 128.42.5.4. Note also that the round-trip time between Palo Alto and Houston is quite regularly about 63 milliseconds.

If you like puzzles, you can dig even further into these results and determine (from line #1) that I have DSL service from PacBell, that at least three communications companies handled parts of the path (sbcglobal, Sprintlink, and Verio) and that the path went through Santa Clara, California (sntc), San Jose (snjsca), Palo Alto (plalca), Dallas, Texas (dllstx) and Houston (hstntx) to reach

Rice. Note that my connection made a loop from Palo Alto to San Jose and back before going on to Houston. This is not uncommon in the Internet, and usually happens when one Internet company "hands its traffic off" to another.

Two more examples of TRACEROUTES

Let's try two more examples before you begin your own work. Suppose I am interested in the Internet path between my house and MIT in Cambridge, Massachusetts. Using the same procedure as before I type TRACERT mit.edu and am rewarded with the response shown below. By examining it I can conclude the following things about the path and the routers along the way ---

* The IP address for the router responding the URL mit.edu is 18.7.21.70.
* It takes 14 hops to reach MIT from my house in Palo Alto.
* The round trip delay from Palo Alto to Cambridge and back is regularly about 87 ms. [This is longer than the round trip time to Houston and back. Why would that be?

```
.>tracert mit.edu

icing route to mit.edu [18.7.21.70]
:r a maximum of 30 hops:

L    15 ms     15 ms     15 ms   adsl-67-118-3-254.dsl.sntc01.pacbell.net [67.118
.254]
?    15 ms     15 ms     15 ms   dist1-vlan50.sntc01.pbi.net [63.203.35.65]
}    14 ms     15 ms     16 ms   bb1-g1-0.sntc01.pbi.net [63.203.35.17]
!    16 ms     15 ms     15 ms   bb2-p6-0.sntc01.sbcglobal.net [151.164.40.170]
;    16 ms     15 ms     15 ms   ex1-p10-0.eqsjca.sbcglobal.net [64.161.1.46]
;    14 ms     15 ms     16 ms   asn174-cogent.eqsjca.sbcglobal.net [151.164.89.2
]
?    15 ms     15 ms     15 ms   p6-0.core01.sjc01.atlas.cogentco.com [154.54.2.2
]
}    18 ms     18 ms     18 ms   p4-0.core01.sfo01.atlas.cogentco.com [66.28.4.93
]
?    61 ms     63 ms     60 ms   p14-0.core01.ord01.atlas.cogentco.com [66.28.4.1
]
]    86 ms     84 ms     84 ms   p14-0.core01.bos01.atlas.cogentco.com [66.28.4.1
]
L    84 ms     85 ms     87 ms   g8.ba21.b002250-1.bos01.atlas.cogentco.com [66.2
.14.210]
?    86 ms     87 ms     87 ms   MIT.demarc.cogentco.com [38.112.2.214]
}    86 ms     86 ms     87 ms   W92-RTR-1-BACKBONE.MIT.EDU [18.168.0.25]
     86 ms     87 ms     87 ms   SOLTICUS-RECKHOMOE.MIT.EDU [18.7.21.70]
```

Figure 8.20 *tracert mit.edu* command output

* The first four hops are very similar to the path from Palo Alto to Rice. [Why would this be?]
* A communications company named cogentco.com handled the path from San Jose to Boston.
* The connection passed through Santa Clara and San Jose again, but also traveled through San Francisco (sfo), Chicago (ord), and Boston (bos). [The three-letter place abbreviations are often known as "airport codes", since they are same as the letters used on your luggage tags. Not all data communications companies use the standard airport codes though, so you often have to be mentally agile to figure out what they are using.]

Now let's go overseas. The last example is an attempt to explore the path from Palo Alto California to the Australian National University in Australian's capital, Canberra. To do this we type:

TRACERT anu.edu.au

and are again rewarded with a successful response. [Warning: You won't always be successful. Some network managements set up their routers so that they will not respond to the probing signals sent out by TRACERT. In this case you will get (no response) or lines of * * *.) In this case we reach Canberra in 15 hops.

You should now be able to analyze this response in the same way that we did for Rice and MIT. The really notable difference is the big jump in round-trip time delay encountered at hop 11 as the path leaves San Francisco.

What would account for this long delay?

```
\>tracert anu.edu.au

icing route to anu.edu.au [150.203.2.43]
:r a maximum of 30 hops:

L    13 ms    15 ms    15 ms   adsl-67-118-3-254.dsl.sntc01.pacbell.net [67.118
.254]
?    14 ms    15 ms    16 ms   dist1-vlan60.sntc01.pbi.net [63.203.51.65]
}    14 ms    15 ms    15 ms   bb1-g2-0.sntc01.pbi.net [63.203.51.1]
┤    16 ms    15 ms    16 ms   bb2-p9-0.sntc01.sbcglobal.net [151.164.40.166]
;    15 ms    15 ms    15 ms   core2-p6-0.crscca.sbcglobal.net [151.164.242.125

;    16 ms    16 ms    15 ms   core1-p1-0.crscca.sbcglobal.net [151.164.241.233

?    16 ms    15 ms    15 ms   core2-p11-0.crsfca.sbcglobal.net [151.164.242.9€

}    15 ms    15 ms    15 ms   bb1-p8-0.crsfca.sbcglobal.net [151.164.243.2]
)    18 ms    18 ms    18 ms   ex1-p13-1.pxpaca.sbcglobal.net [151.164.40.150]

)    17 ms    18 ms    18 ms   151.164.249.50
L   176 ms   177 ms   177 ms   203.208.148.102
?   181 ms   182 ms   182 ms   Gi8-0-0.ci1.optus.net.au [202.139.191.131]
}   183 ms   183 ms   182 ms   ACT-RNO-INT.ci1.optus.net.au [202.139.139.130]
┤   183 ms   182 ms   182 ms   anu-huxley.carno.net.au [203.22.212.66]
;   183 ms   182 ms   182 ms   anumail3.anu.edu.au [150.203.2.43]
```

Figure 8.21 *tracert anu.edu.au* command output

Laboratory Procedure

Steps	Instructions
Step 1:	Pick any three universities in the world, perform a TRACEROUTE to each.
	Q1: Analyze the results to determine the IP address of the destination router, the number of hops to reach it, the average round-trip time delay, the major cities though which the path travels, and, as possible, which companies are involved.
Step 2:	Repeat the first example using the IP address 128.42.5.4.
	Q2: Are the results the same or different? Why?

Steps	Instructions (Continued)
Step 3:	Pick the most interesting example of the three you did earlier and draw the route on a map.
	Q3: If Internet traffic could travel at the speed of light (186,300 miles/sec) around the circumference of the earth, what would be the longest round-trip time you would expect to see?
	Q4: If internet traffic needs to travel through a geosynchronous satellite located 24,000 miles above the earth, what round-trip delay might be expected if a satellite is one of the hops in the path? Given this number, could "Hop 11" in the Australian example be due to a satellite?

Overview Questions

A: Examine your examples for indications of hardware technology, such as references to DSL, dialup modems, T1, and T3 communications links?

B: TRACEROUTE results have common references to three different types of routers - core, boundary, and gateway. What are these different functions? You can find this information in other reference material, but it is also possible to deduce if by just examining enough TRACEROUTE results.

Summary

Through the use of traceroute we have explored the connections that comprise the Internet.

Appendix A: Troubleshooting Guide

The following suggestions are designed to address commonly-asked questions regarding the Infinity Technology Kit. In the event that you encounter issues which are not listed here, please feel free to use either of the following support resources for Infinity Project users:

1. Post a message on the Infinity Project Technology Kit discussion board (http://vab.infinity-project.org/tech_kit_discuss.html). This discussion board is password-protected; please send an email to ipmail@infinity-project.org if you need assistance in accessing the board.

2. Send an email to ipmail@infinity-project.org.

General

Issue	Possible solution(s)
When I start VAB, it won't do anything because it can't find the SPEEDY-33 board	• Make sure that the board is connected to the PC using the USB cable. • Check that the board is connected to exactly the same USB port used during the installation. • Check to see if the green power indicator light on the SPEEDY-33 board is on. If it is not, try connecting the SPEEDY-33 to the PC using a different USB cable.
When I try to open an existing worksheet, I receive a "block not found" error message.	• From the Blocks menu, click "**Auto Build Menu…**" Be sure that the dialog box contains the directory names "C:\Program Files\Hyperception\VABINF" and "C:\Program Files\Hyperception\VABINF\Hierarchy" and click **OK**. • Log on to your computer as an administrator. Open a command prompt window (from the Start menu, click "**Run…**" and type "cmd" without quotes). At the prompt, enter the following command: *cacls "C:\Program Files\Hyperception\VABINF*.*" /t /e /g Everyone:F*
When I try to run my worksheet, I receive the following error message: "Unable to resolve recursion. Worksheet not compiled. Please verify all input connections and recompile."	Make sure that each block input on the worksheet is connected to its appropriate corresponding source. (Hint: to track your connections, click on connection lines using your cursor in setup mode (which resembles the standard arrow-shaped Windows cursor). A pair of white dashed lines will appear on the outside of the selected connection line, which may help you to identify the source and destination of each connection.)
I can't close my worksheet or open a new worksheet.	Make sure you have stopped your worksheet before attempting to close it or open a new worksheet.
My slider only produces integers, but I'd like to see floating-point values.	• Double-click the slider and change the Precision setting to either Float or Double. • Double-click the slider and set the Decimal Places setting to a non-zero value.

Issue	Possible solution(s)
My slider should produce integers, but sometimes it produces an integer that is different from the value shown on the slider.	• Double-click the slider and check to make sure that the Precision value is set to Integer. • Check to make sure that the number of steps is set to associate an exact integer with each position (e.g., number of steps = top value - bottom value + 1 to step by ones). To step by tens instead of ones, make sure that (top value - bottom value) is a multiple of 10 and set number of steps = (top value - bottom value)/10 + 1.

Audio

Issue	Possible solution(s)
I don't hear any sound when I try to run a PC-based audio sheet (audio source connects to the speaker block via a white connection line).	• If you are using external speakers, make sure your speakers are connected to the PC and turned on. • If you are using the internal speaker on your PC, make sure the PC audio volume is turned up to an audible level. • If you are using passive speakers (speakers which are not powered from a wall outlet or batteries), you may need to increase the PC audio volume.
I don't hear any sound when I try to run a DSP-based audio sheet (audio source connects to the speaker block via an orange connection line).	• Make sure your speakers are connected to the "Output" jack of the DSP board and turned on. • To be sure that your DSP board is set up correctly, run Lab 1.0.1 (Testing Your System Mic Cosine) and verify that you are able to hear both the microphone and cosine output. • Increase the amplitude of the audio source(s). • Be sure that all signal generator blocks (such as sine, cosine, square wave, or triangle wave generators) have nonzero frequencies in an audible range. Technically, this range is from 20 Hz to 20,000 Hz, though frequencies below a few hundred Hz or above 10,000 Hz may be difficult to hear on some speakers. • Be sure that all sine and cosine blocks have their Precision values set to Float.
My PC-based audio sounds choppy.	Increase the framesize of the audio source (File Read block, Sine Generator block, etc.).
I'm uploading audio data from the DSP to the PC for viewing, but it's not showing up on the display.	Increase the framesize of your audio source(s).·Connect a Buffer block between the DSP to PC Upload block and your display. (Note: Be sure to use a PC-based Buffer block.)
When I try to add two audio signals on the DSP, I only hear one of them.	Make sure that both audio sources have the same Framesize parameter values.·Check to be sure that neither of the sources has a much higher amplitude than the other.
When I try to run my worksheet, I receive an error message which reads, "Sorry, you have no video capture hardware."	Be sure that your PC camera is connected to your PC and that you have installed its drivers.

Issue	Possible solution(s)
When I try to run my worksheet, I receive an error message which reads, "The system cannot find the file specified."	Double-click the Bitmap Read block and check to see that the desired file name and path are entered in the correct places.
The Image Size block doesn't seem to be producing accurate values.	Double-click the Bitmap Read block and set the Precision value to Integer.
When I try to do mathematical operations on images (such as adding two images, multiplying by a constant, and so on), the result doesn't look right.	Double-click the Bitmap Read block and set the Precision value to Float.

Appendix B: Running MIDI-VAB Labs

What do I need to do to run the MIDI VABs?

1. Open the MIDIBar program (click on 'Start' ® 'Programs' ® 'MIDI-OX' ® 'MIDI BAR').

2. Now click the second button from the left on the MIDI BAR. This lets you select the MCI port that you would want to use for the MIDI file.

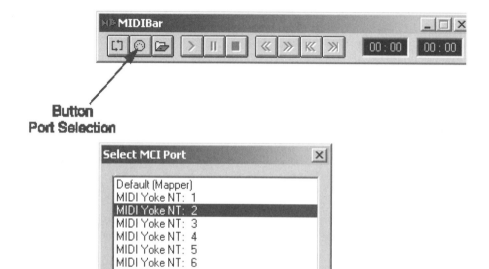

Select any one of the listed MIDI ports. (Note: depending on your operating system, you may have a different number of ports than what is shown in the above diagram. This is just fine, since we usually only deal with one MIDI file at a time.)

3. Now click on the **Open File** button (third from the left on the MIDIBar). This lets you select the MIDI file you want to play. You and your students can download MIDI files from the Internet

by following the instructions above. Select the required file and **PLAY** it by clicking on the **PLAY** button.

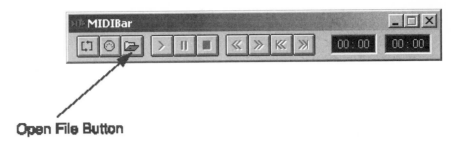

Open File Button

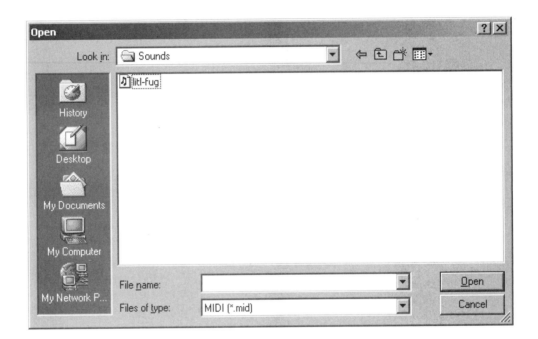

4. Now open any of the VAB designs that play MIDI files. For example, let's try the MIDI Player in Chapter 2 (*L02-05-02 Building the Sinusoidal MIDI Player Test.Lst*). Find and open this file in VAB.

5. Once the MIDI Player worksheet is open, double-click on the **Midi In** block and select the port. This port should be the same as the MCI port you selected on the MIDIBar. After choosing the correct port, click **OK** to close the dialog box.

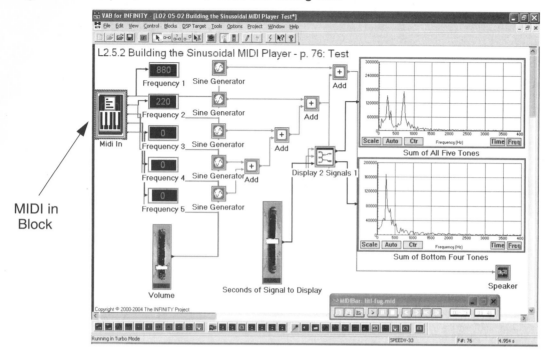

MIDI in
Block

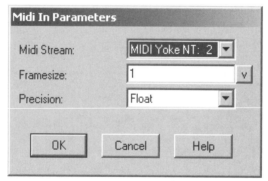

6. Click the **Play** button on the MIDIBar (fourth button from the left).

Play button

7. Click the **Run** button in VAB. You should hear the music from the selected MIDI file.